Interconnection Network Reliability Evaluation

Scrivener Publishing
100 Cummings Center, Suite 541
JBeverly, MA 01915-6106

Performability Engineering Series
Series Editors: Krishna B. Misra (kbmisra@gmail.com)

Scope: A true performance of a product, or system, or service must be judged over the entire life cycle activities connected with design, manufacture, use and disposal in relation to the economics of maximization of dependability, and minimizing its impact on the environment. The concept of performability allows us to take a holistic assessment of performance and provides an aggregate attribute that reflects an entire engineering effort of a product, system, or service designer in achieving dependability and sustainability. Performance should not just be indicative of achieving quality, reliability, maintainability and safety for a product, system, or service, but achieving sustainability as well. The conventional perspective of dependability ignores the environmental impact considerations that accompany the development of products, systems, and services.

However, any industrial activity in creating a product, system, or service is always associated with certain environmental impacts that follow at each phase of development. These considerations have become all the more necessary in the 21st century as the world resources continue to become scarce and the cost of materials and energy keep rising. It is not difficult to visualize that by employing the strategy of dematerialization, minimum energy and minimum waste, while maximizing the yield and developing economically viable and safe processes (clean production and disposal at the end of the life. This is basically the goal of performability effect on the environment during production and disposal at the end of the life. This is basically the goal of performability engineering.

It may be observed that the above-mentioned performance attributes are interrelated and should not be considered in isolation for optimization of performance. Each book in the series should endeavor to include most, if not all, of the attributes of this web of interrelationship and have the objective to help create optimal and sustainable products, systems, and services.

Publishers at Scrivener
Martin Scrivener (martin@scrivenerpublishing.com)
Phillip Carmical (pcarmical@scrivenerpublishing.com)

Interconnection Network Reliability Evaluation

Multistage Layouts

Neeraj Kumar Goyal
Indian Institute of Technology Kharagpur, India
and
S. Rajkumar
Adama Science and Technology University, Ethiopia

Scrivener Publishing

WILEY

This edition first published 2020 by John Wiley & Sons, Inc., 111 River Street, Hoboken, NJ 07030, USA and Scrivener Publishing LLC, 100 Cummings Center, Suite 541J, Beverly, MA 01915, USA

For more information about Scrivener publications please visit www.scrivenerpublishing.com.

Wiley Global Headquarters
111 River Street, Hoboken, NJ 07030, USA

For details of our global editorial offices, customer services, and more information about Wiley products visit us at www.wiley.com.

Library of Congress Cataloging-in-Publication Data

ISBN 978-1-119-62058-7

Cover image: Pixabay.Com
Cover design by Russell Richardson

Set in size of 12pt and Minion Pro by Manila Typesetting Company, Makati, Philippines

Contents

Series Editor Preface

The Figure given below represents the concept of performability as introduced by the editor in 2005 and reflects a holistic view of designing, producing, using, and disposing products, systems or services, which will satisfy not only the basic operational requirements to the best possible extent with minimum cost, but are also sustainable. Sustainability aspect of products, systems and services includes that they consume minimum material and energy in their production, use and disposal and produce minimum waste in order to create minimum environmental impact which is of prime importance and requirement in the 21st Century. In other words, it should represent the holistic performance.

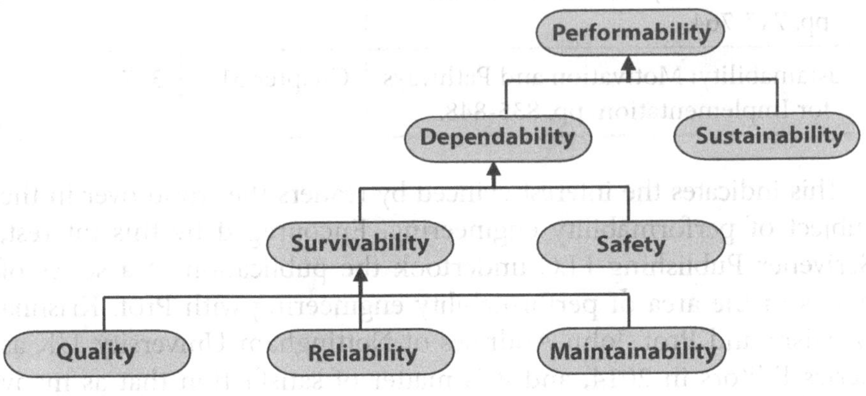

Concept of Performability (Adapted from Fig.1.2 of [1])

In fact, the concept of performability and sustainable development are closely interlinked. To promote the concept to wider section of engineering community, consisting of planners, designers, manufacturers, researchers, technologists and users, the Editor published the first book on performability engineering in 2008 [1]. The book consisted of 76 chapters with 100 contributors and touching on all aspects of performability with the Editor himself contributing 15 chapters. The response to this book has been overwhelming with more than 495,445 chapter downloads to the end of 2019. A German company, ResearchGate, which keeps track of chapters, books etc., read by people the world over, reports that the main chapters authored by the Editor himself had the following statistics as of June 04, 2020:

Title of the Chapter	**Chapter #**	**Total Reads**
Quality Engineering and Management, pp. 157-170.	Chapter 12	18813
Reliability Engineering: A Perspective, pp. 253-288.	Chapter 19	8324
Risk Analysis and Management: An Introduction, pp. 661-674. (Safety)	Chapter 41	1602
Maintenance Engineering and Maintainability: An Introduction, pp. 747-764.	Chapter 46	7477
Sustainability: Motivation and Pathways for Implementation, pp. 835-848.	Chapter 51	377

This indicates the interest evinced by readers the world over in the subject of performability engineering. Encouraged by this interest, Scrivener Publishing LLC undertook the publication of a series of books in the area of performability engineering with Prof. Krishna B. Misra and Prof. John Andrews of Nottingham University, UK as Series Editors in 2014, and it is matter of satisfaction that as many as 10 books have been published under this series so far. The books published under this series have solicited good number of citations in spite of the fact that much time has not elapsed after their publication.

As an Editor of this series, I solicit manuscript of books from academicians, engineers and researchers in any area of Performability Engineering. For details, please contact the Series Editor.

Krishna B. Misra, Series Editor
kbmisra@ramsconsultants.org
Web: www.ramsconsutants.org
June 2020

[1] Misra, Krishna B. (Editor), Handbook of Performability Engineering, Springer, London, 2008.

Preface

Network reliability evaluation for complex communication networks is a time-consuming task. Complex networks such as computer communication networks, telecommunication networks, transport networks, etc. are an integral part of our daily life. Complex networks interconnect multiple processing units in distributed systems. Performance of such networks can be measured by evaluating the reliability index, which represents the probability that the network operates satisfactorily for a given period of time when used under stated operating conditions. A network designer has to choose suitable fault-tolerant and highly reliable network architecture to avoid performance degradation and carry out multiple communication tasks concurrently.

Parallel computers with multiple processors have emerged to meet energy-efficient performance demands of current and future challenging computing applications. These multiprocessor systems need interconnection networks for connecting processors and memory modules. Advances in parallel and distributed computing have made interconnection networks a potential networking alternative to meet the growing demands of high-performance computing applications. These networks need to perform continuously without repair for long periods of time. Therefore, these need to be reliable and fault-tolerant as these networks are quite complex and big in size. Failure of its components should not lead to complete network failure as it may be catastrophic.

The main objective of this book is to device approaches for reliability modeling and evaluation of such complex networks. Such evaluation helps to understand which network can give us better reliability by their design. New designs of fault-tolerant interconnection network layouts are proposed, which are capable of providing high reliability through path redundancy and fault tolerance through reduction of common elements in paths. This book covers the reliability evaluation of various network topologies considering multiple reliability performance parameters (two terminal reliability, broadcast reliability, all terminal reliability, and multiple sources to multiple destinations reliability).

For decades, work on network reliability evaluation has been going on. Our S.C. School of Quality and Reliability, earlier known as Reliability Engineering Centre, has significantly contributed to this area of research during the 1980s and onwards. The school has been involved in developing models for reliability evaluation of complex networks but static and ad hoc. The authors of this paper have significant contributions in this area.

Interconnection networks pose another challenge to reliability evaluation due to their complex and large architectures. There are very few in this area. This subject matter has not been much addressed in easy-to-follow book materials. This book tries to bring the aspects of latest research conducted at S.C. School of Quality and Reliability to have better reach to audiences. This is the first edition of the book, so there are many improvements possible. The authors can be provided with feedback to improve the contents of the book in the possible next edition.

This book first introduces the basic concepts of network reliability and the popular interconnection network architectures in the first two chapters. Then it discusses approaches used for reliability evaluation of such a network in Chapters 3, 4, and 5. Some new novel architectures providing better reliability and fault tolerance while being cost effective are presented in Chapter 6.

The authors would like to thank everyone (faculty, students, and staff) from S.C. School of Quality and Reliability, IIT Kharagpur, India, for their encouragement, suggestions, and support while carrying this analysis. We would also like to thank Prof. K.B. Mishra and Martin Scrivener for their guidance and support in writing this book.

Dr. Neeraj Kumar Goyal and Dr. S. Rajkumar
August 2020

1
Introduction

1.1 Introduction

In recent years' human beings have become largely dependent on communication networks, such as computer communication networks, telecommunication networks, mobile switching networks etc., for their day-to-day activities [1]. In today's world, humans and critical machines depend on these communication networks to work properly. Failure of these communication networks can result in situations where people may find themselves isolated, helpless and exposed to hazards. It is fact that every component or system can fail and its failure probability increases with size and complexity.

Therefore, it is essential to compute and assure the reliability of these networks, which are growing and becoming complex. Reliability modeling and computation is necessary for reliability and safety assurance of these networks [2]. It also helps to identify weak links. These weak links can be improved cost effectively using reliability design techniques. Recent developments in communication hardware industry has resulted in increasingly reliable and non-repairable (need to be replaced when got bad 1) network components. However new designs involve new components, which tend to be less reliable. A good network design [3] involving fault tolerance and redundancies can deliver better system reliability at lesser cost. This allows new designs to be released faster and works reliably even when components are not mature enough from reliability point of view.

The computation of reliability measures [4] for a large and complex communication network, up to the desired level of accuracy, is time consuming, complex and costly. It has not been realistic to

Neeraj Kumar Goyal and S. Rajkumar. Interconnection Network Reliability Evaluation: Multistage Layouts, (1–10) © 2020 Scrivener Publishing LLC

model and compute the reliability of real-life communication networks, which are quite large, using desktop computer due to large execution time and high memory requirements. Such computations are usually performed on high-end processors for critical systems only. Reliability professionals and researchers have carried out a lot of research and developed techniques to minimize these efforts and develop a practical tool for all the communication network designers [4–6].

This book presents novel and efficient tools, techniques and approaches for reliability evaluation, reliability analysis, and design of reliable communication networks using graph theoretic concepts.

1.2 Network Reliability Measures

Earlier attempts to measure network reliability belong to two distinct classes: deterministic and probabilistic [1, 2]. The deterministic measures assumed that the network is subjected to a destructive force with complete knowledge of the network topology. The reliability is measured in terms of the least amount of damage required to make the network inoperative.

Deterministic measures thus provide simple bounds on the reliability of the network, since they are often measured for the network's worst-case environment. For example, in the terminal pair reliability problem, two deterministic measures of reliability are:

1. The minimum number of edges that must be destroyed or removed to disrupt the communication between the specified nodes (s and t), which is simply the number of edges in a minimum cardinality cutset, and
2. The minimum number of nodes that must be destroyed or removed to disrupt the communication between the specified nodes (s and t).

Both of these measures are computable in polynomial time. However, one of the main problems with deterministic measures is these give rise to some counter intuitive notions of network reliability. For example, consider the networks shown in Figure 1.1.

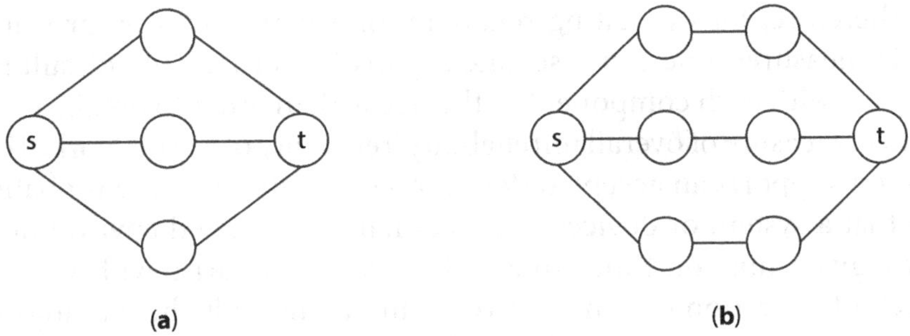

Figure 1.1 Example networks for deterministic reliability measurement.

According to second deterministic measure of the graph's reliability, the graphs of Figure 1.1 (a) and Figure 1.1 (b) are equally reliable since both of these require minimum three nodes to be destroyed for breaking the s-f node connectivity.

However intuitively one can easily find out that graph (a) is the more reliable among the two. The same problem arises when the cardinality of a minimum (s, t)-cut set is used as a measure of unreliability. Consider the graphs shown in Figure 1.2. Both graphs (a) and (b) have a minimum cardinality for (s, t) cut of size one, which implies both networks are equally reliable.

This clearly shows that deterministic measures of reliability are insufficient to correctly relate network components used in network layout with network reliability. Moreover, failure of network components is probabilistic in nature therefore only probabilistic measures can define system reliability appropriately.

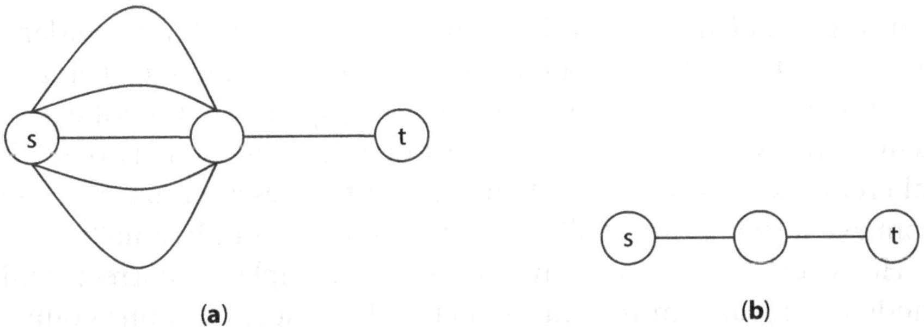

Figure 1.2 Another set of example networks for deterministic reliability measurement.

Therefore, for evaluating reliability of a network using probabilistic measures, one can associate a statistical probability of failure/success with each component of the network in order to obtain a statistical measure of overall unreliability/reliability of the network. This notion supports an accepted definition of reliability as the probability that a system or device is operational under stated environment for a given mission time. To avoid conflicts that arise with various levels of operation within a network's hierarchy, only the topology of the network is considered. This allows a network to be modeled by a graph where the nodes of the graph represent the communication centers and communication links are represented by its edges.

1.3 The Probabilistic Graph Model

Communication networks are generally modeled using network graph [3]. The network graph G (V,E) consists of a set *V* of *n* number of nodes (or vertices) and a set *E* of *l* number of edges (or links). For reliability evaluation, probabilistic graph is used which takes these sets *V* and *E* of nodes and links as random variables. In probabilistic graph of communication networks, nodes represent the computers/switches/transceivers/routers and edges represent various types of communication links connecting these nodes. For reliability analysis, graphical models of networks are considered to be simple, efficient and effective.

Probabilistic graph models are developed and presented in this book. Depending on the state (working or failed) of nodes (or vertices) and/or links (or edges), the network can be considered either working or failed. A general assumption of statistical independence among nodes and links failures is followed throughout. It implies that the probability of a link or node being operational is not dependent of the states of the other links or nodes in the network. The inherent assumption here is that the link failures are caused by random events which affect all network components individually.

However, this assumption may not be completely correct while modeling a real communication network as more than one component in a particular area may fail due to natural causes such as a major storm or an earthquake. In such cases, dependency analysis and common cause failure modelling can be used over the analysis performed with assumption of statistical independence. This assumption is often made because of difficulties in obtaining information about the dependencies of link failures and increased modeling and computational rigor. In fact, such dependencies may not be known. Thus, without the assumption of statistical independence the problem becomes much more difficult to solve.

Depending on the connectivity objective of nodes [4–6], the network reliability evaluation problem can be sub-divided into following different cases:

1. Two terminal or terminal pair reliability (TPR) problems: The most common communication operation is to send messages from a source node *s* to a terminal node *t*. The terminal pair reliability of a network is defined as the probability of having at least one operational path between the nodes *s* and *t*. In case of directed networks, it is usually called (s,t) connectedness.
2. Global or all terminal reliability (ATR) problems: The all terminal reliability of a network is defined as the probability that for every node pair (Ni,Nj) there exist an operational path to connect them; or equivalently, the probability that there exist a working spanning tree. In the directed case, all terminal reliability is the probability that the directed graph contains at least a spanning tree rooted at the source node *s*.
3. K-terminal reliability (KTR) problems: The k-terminal reliability ensures that a specified set of k-nodes of the network are able to communicate with each other and it is defined as the probability that a path exists between every pair of nodes belonging to the specified set of k nodes of the network.

Generally, communication network performance is defined not only by the connectivity between nodes but also by the minimum capacity it can transfer between the nodes. The reliability measure considering both capacity and connectivity, as essential performance criterion, is known as capacity related reliability (CRR). It is defined as the probability that required amount of flow is transferred from source node *s* to terminal node *t*. Evaluation of above network reliability measures (indices) has attracted a lot of attention from researchers and many approaches have been developed so far. Next section presents a brief summary of these approaches.

1.4 Approaches for Network Reliability Evaluation

Misra and Rao [4] developed signal flow graphs: a development recognized as a significant step forward in the evaluation of network reliability. After this, a number of algorithms, techniques and approaches have been suggested in the literature. In fact, today, the use of graph theory has become inseparable from network reliability evaluation. Available literature on reliability evaluation of communication networks, considering only connectivity as performance criterion, can broadly be classified into two paradigms, viz.:

Path sets or Cut sets approaches (POC) paradigm:
These use pathsets or cutsets as the starting point for TPR problems, spanning trees for ATR problems and k-trees (k-minimal cutsets) for KTR problems. In this book, the terms path sets and trees are represented by a single term path sets as approaches used for generating them are similar. From the context, whether it is pathset or spanning tree or k-tree can easily be understood. For example, when the context is TPR problem then it is pathset, when context is ATR it is spanning tree and when the context is KTR it is k-tree.

Reliability evaluation is generally achieved by enumerating pathsets or cutsets of the network. A pathset is defined to be a set of minimal paths connecting source and destination node. A path is set of components whose functioning ensures that the system functions. A cutset is defined to be a set of minimal cuts that disconnects the source and destination nodes. A cut is a set of components whose failure will result in system failure. A minimal path/minimal cut is a path/cut such that no proper subset of minimal path or minimal cut is a minimal path/minimal cut. In other words, if any element is removed from the set (minimal path/minimal cut) then it no longer remains a path or cut.

Non-Path sets or Cut sets approaches (NPOC) paradigm:
These approaches do not use path sets or cut sets. These apply reduction or decomposition or transformation or a combination of these approaches to solve the reliability evaluation problems. The non-path (or cut) set approaches may be categorized as following: 1) State enumeration methods 2) Topological methods 3) Transformation methods and 4) Decomposition (Factoring theorem) methods.

1.5 Motivation and Summary

Main objective of this book is to design new fault tolerant Interconnection network layouts capable of path redundancy among dynamic failures. New INs designs have been proposed and their observed results are found promising when compared with some of the earlier networks.

A summary of problem wise contributions is discussed next:

1. *INs Topology Review* (**Chapter 2**)
   Presents an extensive survey of the existing INs topological hardware aspects based on its unexplored Taxonomy of INs performance metrics. In this chapter detailed literature review on interconnection networks has been performed, and an exhaustive taxonomy has been proposed based on the literature review. The chapter also presents advantages and disadvantages of interconnection networks along with a summary of topological characteristics and fault tolerance information of the surveyed interconnection networks in the form of Table.
2. *MIN Reliability Evaluation Techniques Review* (**Chapter 3**)
   Presents an overall view of the different reliability measures and their importance in evaluating the reliability can be reviewed with examples. A comprehensive review of the various approaches present for evaluation of different reliability measures is also presented. Based on the literature survey a few shortcomings are identified.

3. *Terminal Reliability Analysis of Existing MIN Layouts* (**Chapter 4**)

   Networks such as telecommunication, transportation, power systems, integrated circuitry and computer communication systems are large and it becomes imperative for economic reasons and safety that these have high reliability. Reliability evaluation of such complex systems has not yet been perfected and there is a lot more that needs to be done. System reliability can be measured in terms of various paradigms such as *two terminal*, *all-terminal* and *broadcast* reliabilities.

   In this chapter, terminal reliability values of various recent and industrial used MINs topologies are evaluated and compared using existing simple and efficient path-set enumeration method. Path set and cut set techniques are the most widely used techniques for terminal pair reliability evaluation of communication networks as these techniques give a compact reliability expression in sum of disjoint product (SDP) form.

   These techniques first enumerate minimal path (or cut) set for the network and then evaluate reliability (or unreliability) expressions in the SDP form. Since enumeration of cut set also requires knowledge of basic paths of the network, the path set enumeration becomes a necessity for these techniques. Available disjoint and redundant paths of various existing MINs are evaluated by a simple and efficient path-set enumeration method based on hybrid multi-variable inversion (MVI) algorithm, which gives accurate reliability values. Present reliability evaluation paves the way of research focus on developing higher disjoint, reliable and highly fault-tolerant MINs than the existing topologies.

4. *Comprehensive MIN Reliability Paradigms Evaluation* (**Chapter 5**)

   This chapter has attempted to evaluate comprehensive reliability paradigms of various recent MINs topologies available from literature using path tracing algorithm.

With increasing number of input and output nodes in supercomputer environment reliability evaluation of multi-cast nodes is mandatory. In other words, a network has to transmit a signal or commodity between different pairs of nodes of a network simultaneously. Communication between multiple-sources multiple-destinations offers several advantages like increased performance, improved reliability and decreased costs through resource sharing.

So we have extended the traditional reliability evaluation to include a new reliability measure as multi-source multi-terminal reliability (MSMT). This chapter also evaluate MIN traditional reliability parameters (Two/broadcast/all terminal reliability) using conceptually simple and computationally efficient path tracing algorithm.

5. *Dynamic Tolerant and Reliable Four Disjoint MIN Layouts* (**Chapter 6**)
This chapter proposes two new fault tolerant and highly reliable multistage interconnection network layouts. Proposed 4-disjoint paths multistage interconnection networks (4DMINs) provide four disjoint paths between any source and destination node pair and provide dynamic rerouting between intermediate stages. Minimal paths of these 4DMIN have been evaluated and MINs performance evaluation has been done by calculating terminal reliability at assumed different switch reliabilities using existing approaches of sum of disjoint products method.

Then a new layout design of fault tolerant reliable interconnection networks (RINs) is proposed. Proposed INs dcsigned with highly adaptive reliable multipath dynamic routing behavior and fault tolerant capability in all the stages with suitable routing techniques. Proposed design can withstand dynamic switch failures in all the stages and provide more redundant paths to tackle such with higher reliability. They provide higher reliability and outperform other MINs.

Recently, research has been undertaken to improve interconnection networks for incoming the next generation of large parallel and distributed systems. Therefore, research focused on designing highly reliable and fault tolerant INs because they provide faster intercommunication capability. This book attempted towards the evaluation of reliability of interconnection networks and concentrates more on reliability evaluation of computer communication networks. But the methods have a good potential to explore new reliability evaluations using the algorithm presented in this work in other fields of interconnection networks systems such as water distribution systems, banking industry, and power distribution systems and any real time interconnection networks.

2
Interconnection Networks

2.1 Interconnection Networks Classification

In today's era, high computing speed and power efficiency is needed in many application areas [7–10] like telecommunication, mathematical and scientific applications, industrial automation, geographical simulation, artificial intelligence, genetic engineering, biological simulation and system simulations etc. These applications require parallel computers with multiple processors to meet performance demands of current and future challenging computing applications besides being energy-efficient. Multiprocessor systems are emerging system architectures to address power consumption and performance scaling issues by integrating thousands of processors.

These multi-processor systems need interconnection networks for connecting processors and memory modules. Interconnection networks have been studied for more than last two decades and a solid foundation of design techniques has been described [12, 13]. Advances in parallel and distributed computing have made interconnection networks a desirable networking alternative to meet the growing demands of high-performance computing.

An interconnection network [15–18] is made of three fundamental blocks: logic, memory, and communication. Interconnection networks are primarily utilized in tightly coupled multiprocessor systems and large-scale parallel/distributed systems for the purpose of broadband switching and inter-processor communications. Interconnection network design is a key element in developing large parallel computing systems. These specific-purpose networks have been applied in real time environments efficiently. Advanced computer architectures utilize parallelism *via* multiple processing units. The responsibility of interconnecting these units is with interconnection networks (INs)

Neeraj Kumar Goyal and S. Rajkumar. Interconnection Network Reliability Evaluation: Multistage Layouts, (11–62) © 2020 Scrivener Publishing LLC

[19–22]. The processing units connect with each other through IN to solve complex problems. Interconnection networks offer an attractive and cost-effective solution to communication and interconnection between system components. Therefore, design of an efficient interconnection network is an essential task for construction of high-performance multiprocessor and microelectronic systems [23].

A significant part of the early work on interconnection networks [24–28] was spurred by the needs of the communication industry, especially in the context of telephone switching to provide fast and efficient communication, but with the growth of the computer industry, various applications for interconnection networks emerged within computers as well. The first amongst these was sorting of numbers. Soon, interest in parallel processing grew and substantial number of interconnection networks was proposed for interconnecting processors and memory modules.

The communication infrastructure interconnection network design essentially points to the design of underlying hardware acting as the backbone for the communication network. Numerous approaches of network interconnection have been proposed, ranging from a single bus to full mesh connected architectures. Interconnection networks provide a communication medium in multi-processor system as they interconnect a number of processors and memory modules.

Interconnection networks [19] topologies can be classified into shared medium, direct, indirect and hybrid interconnection networks. Local area networks and backplane buses are shared-medium interconnection networks where transmission medium is shared by all communicating devices. Hybrid Networks is made up of a combination of direct and indirect topologies. Line/Ring, mesh, torus, tree, fat tree, k-ary n-cubes and hypercube are called as static (direct) interconnection networks where the nodes directly communicate through point-to-point links with each other. Static interconnection networks are widely used for system on chip, high performance parallel and distributed systems [7] like Cosmic Cube, J-Machine, Octagon, Mark III, Series 2010, Proteo, iPSC, FPS T Series, SOCBUS, Rediflow and NCUBE.

The irregular topologies like shared bus, crossbar, Multistage interconnection networks (MIN) belongs to dynamic (indirect) interconnection networks in which communication among the nodes takes place through switching network. These indirect INs are mainly used for switching and routing the dynamic nodes in parallel computing applications.

Shared medium arbitrated bus is the commonly used communication backbone interconnection architectures. Although this architecture has the advantages of simple topology, extensibility, and low area cost, a shared bus allows only one communication at a time which may block all other buses in the hierarchy.

Bus based communication does not scale the system performance with the number of cores attached and full connectivity is too expensive. The main strength of a common bus is that it is simple to implement but on the other hand a single common bus system is restricted to one processor transferring data at a time. Some of the notable interconnections [7] of Encore Multimax, IBM Cell [8], AMD Hyper Transport, Sequent Balance and Intel QuickPath used this topology.

Crossbar networks are one of the most efficient interconnection networks. A centralized crossbar switch overcomes some of the limitations of the buses. The connections through point-to-point links operated at very high speed outperform the shared bus. The crossbar switch possesses complete connectivity with respect to the memory modules. The cross-bar can realize a communication from any input port to any output port. Nevertheless, crossbar networks are highly expensive hardware interconnection architectures. YSE [7–10], Alliant FX/8, C.MMP, Sun Niagara and UltraSPARC T2 interconnect are common examples.

Multistage interconnection networks (MINs) have been extensively studied for parallel and distributed systems [19] because of their scalability and dynamic configuration. MINs have emerged as the viable alternative for the design of modular and scalable communication architectures. Implementing MINs offers considerable advantages. These are inexpensive in terms of cost of system layout. As crossbar networks of large sizes are cost prohibitive, MINs have been developed that provide similar functioning at lesser cost.

MINs development [7–19] has been an area of great interest to computers and communication researchers. They provide higher performance than shared bus and lower switching cost than crossbar networks as presented in Table 2.1.

Table 2.1 Dynamic networks topological features.

	Bus	Crossbar	Multistage
Speed	Low	High	High
Cost	Low	High	Moderate
Reliability	Low	High	High
Configurability	High	Low	Moderate
Complexity	Low	High	Moderate

2.2 Multistage Interconnection Networks (MINs)

From Table 2.1, we can deduce that multistage interconnection network (MIN) [19–23] act as an intermediate performance when compared with shared bus and crossbar for large multiprocessor systems. MINs introduced efficient and flexible communication capability for parallel machines. General structure of MIN is shown in Figure 2.1.

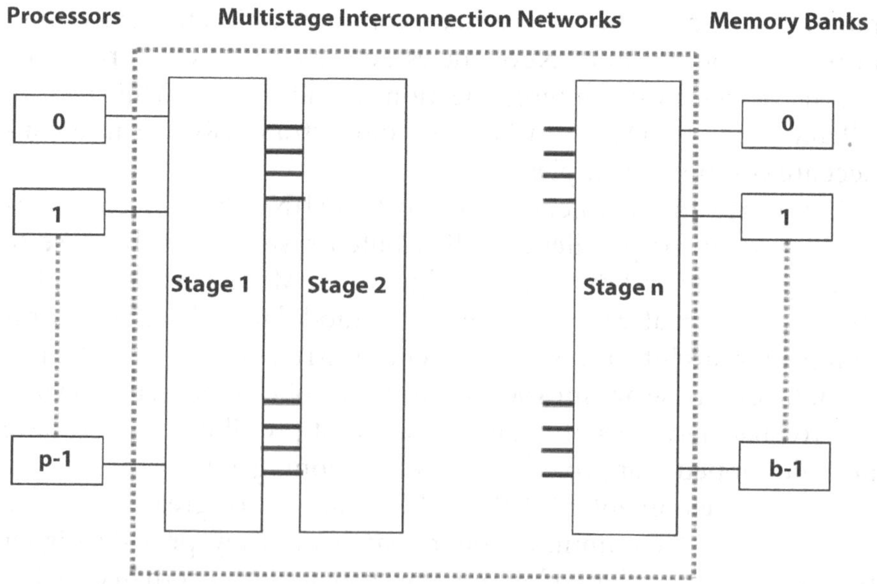

Figure 2.1 General topology of multistage interconnection network (MIN).

An advantage of these MINs over static networks is that they can dynamically setup communication between processors, based on parallel programming requirements. Multistage Interconnection Networks (MINs) [23] are widely used in parallel and distributed systems, communication networks, optical networks and modern embedded systems involving large number of interconnected processor or processor-memory modules. With advancement in research field of multiprocessing [27–30], these systems rely on fault tolerant and highly reliable hardware communication infrastructure. Reliable operation of multistage interconnection networks (MINs) is essential to meet overall system performance requirements.

2.3 Research Issues in MIN Design

The key design dimensions for interconnection networks (Figure 2.2) are classified into three paradigms.

1. Interconnection infrastructure
2. Interconnection management
3. Interconnection evaluation

The interconnection infrastructure plays a central role in determining the overall performance of a multicomputer system. It requires focus on designing best interconnection infrastructure based on application demands, technology supports and cost effectiveness. Modelling of the MIN flow of control has been effectively carried out using high-level Petri net by Chaki [11].

Choice of a particular interconnection network depends upon:

- Topology of Interconnection network
- Router and Buffer architecture design
- Flow control protocol design
- Switching techniques design

Second dimension of research deals with interconnection management paradigm. It is the process of measuring and controlling the information flow (traffic/packets) on a network to avoid exceeding

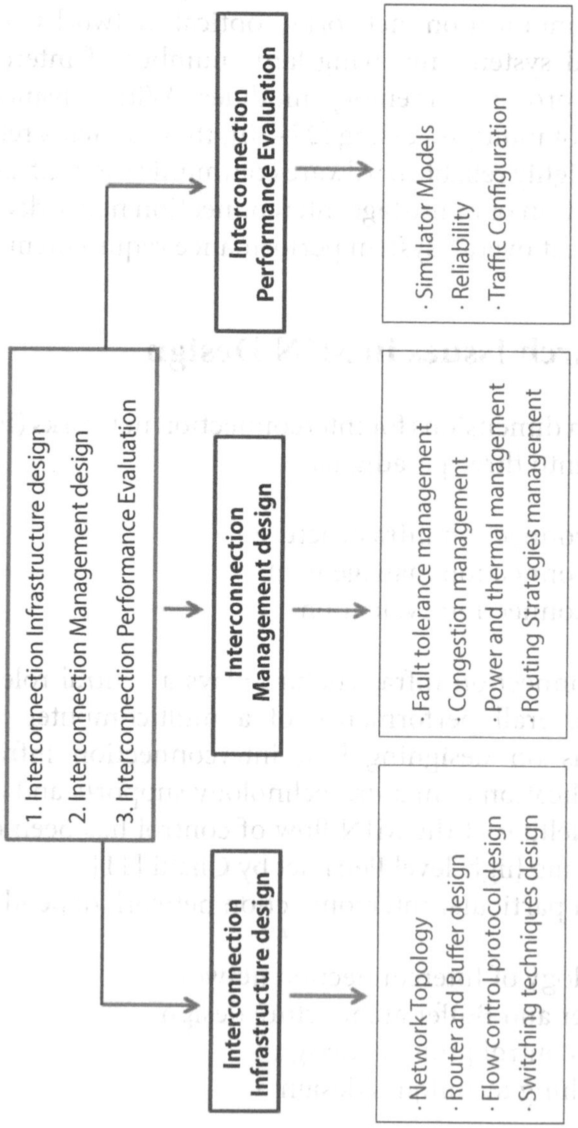

Figure 2.2 Key research issues of multistage interconnection networks.

its maximum capacity, which would result in network congestion and poor performance.

Management aspects help in reducing the overall energy consumption [26], reducing operating costs and prevent faults. Routing strategies [17, 18] will manage transmission of packets of data from source to destination using routing tags which define the route of any message transfer through the switching network.

Various management aspects are:

- Routing Strategies management
- Fault tolerance management
- Congestion management
- Power and thermal management

Performance of a parallel system depends significantly on the effectiveness of the interconnection network. The performance of MINs during design is usually determined using simulation or mathematical models. Third paradigm addresses measurement of interconnection network performance in terms of design metrics.

Performance aspects can be evaluated through:

- Network Simulators
- Interconnection reliability modeling
- Traffic Analysis

A Network Simulator [19] predicts behavior of any network, without employing actual/real network. Simulator tools are used to debug and test MIN to evaluate various performance metrics. In this process, a proper choice of the network performance parameters for the simulation model is essential. There are several available simulators in the literature for featuring the MINs performance in terms of routing, multicasting, mobility simulations, protocol studies, collision detection and reliability.

Veglis [22] used MathCAD and MathConnex to evaluate bandwidth of MINs by considering the various input parameters like number of switches, number of processors, number of memories, number of shared buses, number of stages and number if I/O ports in various SEs. Multistage interconnection network simulator has been designed for the simulation of crossbar and MINs by MINSimulate [20]. The entire simulator was designed in C++ language. When network size increases, this simulator showed limitations giving only approximate results.

CINSIM simulator [21] was developed in C++ for the simulation of component-based interconnection network. This simulator has limited scope to evaluate source or destination throughput, the delay times and the buffer queue sizes. However, the available simulators do not provide accurate results. The results provided differ from that of the real-world scenarios because the simplifying assumptions considered for building the simulators are fail to consider practical behavior. A traffic model attempts to mimic the real movements of the traffic patterns in the nodes influencing the performance characteristics of the multistage interconnection networks.

Designing a reliable network [23–30] is critical for achieving a system that is robust to unexpected failures. Increased reliability would be helpful for several application services. Network reliability is generally used as a measure of robustness of a MIN. Performance of MINs can be measured by evaluating the reliability index, which represents the probability that the network operates satisfactorily for a given period of time when used under stated operating conditions.

Network reliability depends not only on the reliability of network components but also on network topology. The failure of the MIN may occur due to variety of reasons such as failure of a node and/or link, poor choice of network topology, poor fault tolerance and improper routing strategies. Intuitively, MINs are more desirable [27–30] if they are:

1. Efficient
2. Scalable
3. Regular (i.e. easy to implement)
4. Expandable (i.e. highly modular)
5. Interoperable
6. Highly Adaptable
7. Reliable and
8. Fault Tolerant.

Potentially large MINs may comprise thousands or even tens of thousands of processor nodes. It is true that performance suffers as the number of devices grows which makes routing, managing of large networks quite difficult. The scalability is affected because of changing topology, lack of fault tolerance and poor reliability. Interoperability is a property of a MIN system, whose interfaces are completely understood to work with other product/system without any restricted access or implementation. Interoperability is one of the key requirements in almost all areas of application of distributed networks. Interoperability is a necessary parameter for increasing the network coverage and ensuring reliable access. Adaptability is a crucial factor in system development, as it allows systems to cope with increasing complexity and task variability.

Clearly no single interconnection network is best on all criterions [19]. Key research challenge in MINs is to increase the efficiency of information transfer with a reliable design [23], enhancing the use of such networks to different application areas, and work under harsh environmental conditions.

2.4 Some Existing MINs Implementations

MIN is an active research field with many practical applications in industry as it is expected to be an efficient communication backbone of multi-core based parallel computing. The applicability of MINs in digital communication systems with the appearance of Myrinet was adopted in 1995. Few examples of parallel computers with static network [7] are: Cray T3E, Cray T3D (1024 nodes), Intel Cavallino, SGISPISER, MIT M-Machine and Intel iPSC-2 Hypercube.

Some of the existing usage and examples of multiprocessor system [7] that implement MIN are Ultracomputer, NEC Cenju-3 and Cenju-4 IBM RP3, Myricom Myrinet, Thinking machines CM-5, Inmos C104, IBM SP2 and Tandem ServerNet, CARY X-MP series, IBMSP, omega array-processor, flip STARAN, Quadrics (HP Alpha server clusters), BBN Monarch, SGI Altix, Atlas (LLNL), SPIN (UPMC/LIP6), IBM ASCI White SP Power3, ASCI White, IBM Power5, Rediflow, ASCI Blue Mountain, PC Clusters (QsNet,

Autonet), Ranger (TACC), Cedar, IBM ASC Purple pSeries 575 [Federation], HEP, SGI Altix UV, ASCI Blue Horizon, IBM RS6K SP, Sun UltraSPARC T1 processor, Monsoon, TRAC, Meiko CS-2, GF-11, Intel Thunder Itanium2 Tiger4, Thunderbird (SNL) and CHiC (TUC) [7].

2.5 Review of Topological Fault Tolerance

A large number of research papers and reviews [19, 66–71, 79, 80, 123] on multistage interconnection networks exist in literature. These papers address diverse set of performance criterion for designing best architectural MIN interconnect for next generation parallel and distributed systems (i.e., multiprocessors, multi-computers and multi-cores).

Network performance criterions such as throughput, bandwidth, blocking probability and network delay have been adequately analyzed in previous works. Reliability can be considered as another crucial network performance parameter. It should be noted that improvements in reliability [27–30] can improve some other performance parameters such as path length, routing, throughput and blocking. However, there has not been a coherent reliability analysis of these networks so far. Another imperative performance measure of MIN is related to the cost effectiveness of the hardware and scalable MIN design for interconnection.

MINs can be analyzed considering various performance aspects or issues. Some important performance metrics found in the literature [19, 79, 80] include: fault tolerance, permuting ability, reliability, message delay characteristics, throughput, network complexity, and cost-effectiveness. In general, it is difficult to select the "important/fixed" metrics. Their choice will result in different design selection.

The choice of metrics may vary with applications and can be given weights accordingly. From perspectives of near-real scenarios, fault-tolerance and reliability plays a dominant role in design of interconnection networks for large scale systems. Therefore, we have analyzed both these metrics along with cost for comparison of MINs.

Various MIN topologies have been proposed in the last twenty years for large scale parallel processing systems. MINs can be classified on the basis of general characteristics as shown in Figure 2.3.

MINs are comprehensively analyzed in terms of number of redundant and disjoint paths considering topological elements. Topological reviews based on disjoint paths are unprecedented in contrast to many previous works. So far there has not been a clear and logical analysis of the disjoint paths information in these networks. Recent researches belong to increasing number of disjoint paths. This chapter reviews prior study on multistage interconnection networks based on its unexplored Taxonomy (Figure 2.4) of MINs topology factors.

Most of the discussion in MIN design is centered towards fault tolerant capability of designs. The basic idea for the fault-tolerance of the MIN is to create redundant (alternative) paths between any source - destination (S-D) node pair, so that redundant paths can been used when any switch gets failed/busy.

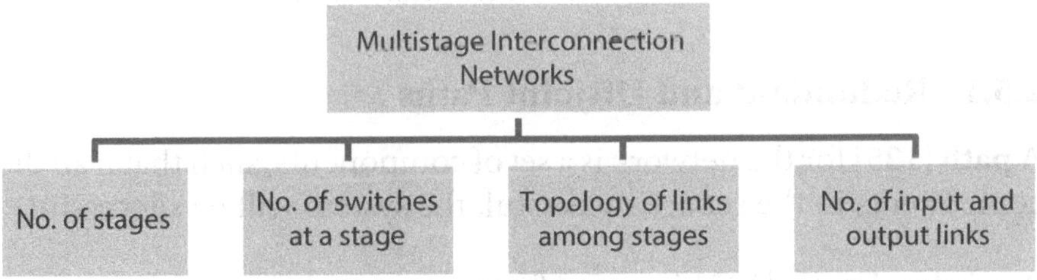

Figure 2.3 MIN general characteristics classification.

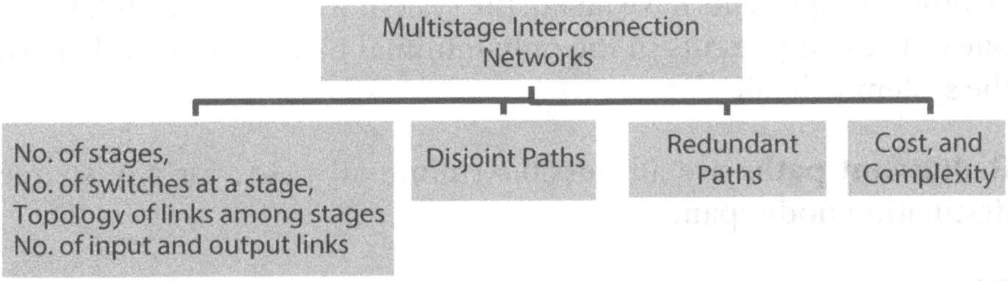

Figure 2.4 Taxonomy of MIN topology.

Fault tolerance in multistage interconnection network [34–40, 45–55, 60–75] can be achieved by many methods introduced by many researchers such as

- Increasing disjoint paths
- Providing dynamic rerouting between the stages
- Incorporating multiple copies of a basic network
- Increasing number of stages
- Increasing size of switches
- Providing multiple links between stages
- Combining switching elements
- Changing the interconnection patterns and
- Backtracking

In general, fault tolerance is enhanced by increasing disjoint paths between every S-D node pair and employing dynamic rerouting between them. Recent literature aimed to design fault tolerant MINs based on both disjoint paths [49, 51, 53–55] and dynamic rerouting capabilities. Fault-tolerance is analyzed in terms of number of redundant and disjoint paths in recent literature.

2.5.1 Redundant and Disjoint Paths

A **path** [125] for the network is a set of components, such that if all the components in the set are successful, the system will be successful.

A **minimal path** [125] is a set of components that comprise a path, but the removal of anyone component from it will make the resulting set not to be a path.

If all the components in any one minimal path are successful while all other components have failed, the system will be successful. If any one of the components in the only minimal path subsequently fails, the system will fail.

Redundant paths are all possible (minimal) paths for any source destination nodes pair.

Disjoint paths: Redundant paths which do not have any element in common other than source and destination nodes.

We can understand the importance of disjoint paths information in MINs through preceding cases and examples.

Case 1: MIN aborts if the common node fails.
Redundant paths: All possible paths for any source destination nodes pairs (1 to 5). For the network in Figure 2.5, we get two paths (**1-2-4-5** and **1-3-4-5**).

Disjoint paths: The disjoint paths are the paths which do not have any common switch other than switches connected to source destination nodes.

We get only one disjoint path (either **1-2-4-5** or **1-3-4-5**). Here, if common switch 4 is failed then communication between S-D node pair is lost.

From Table 2.2, we can deduce that MINs like replicated Benes [67], parallel Benes [67], partial chained GIN [47] and fully chained GIN [47] provides 8, 16, 18 and 64 redundant paths respectively. But they provide only two disjoint paths. Thus, failure of these common elements will abort MIN irrespective of abundant redundant paths.

Case 2: MINs dispatch successfully if they provide more redundant and disjoint paths.
Redundant paths: We get three redundant paths **(1-2-4-6, 1-2-5-6** and **1-3-5-6)** for the network in Figure 2.6.

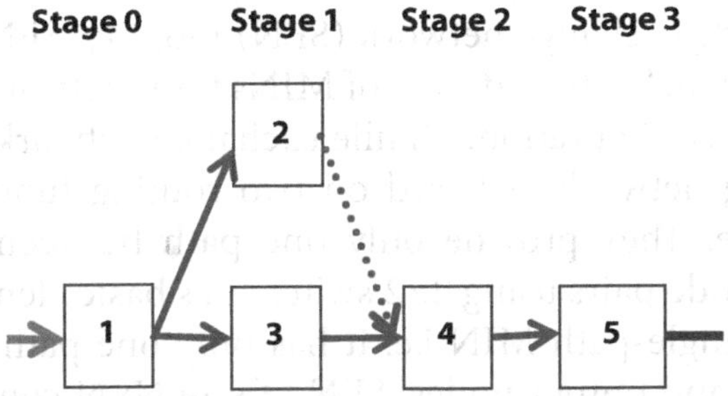

Figure 2.5 Network showing one disjoint path and two redundant paths for S-D node pair (0, 5).

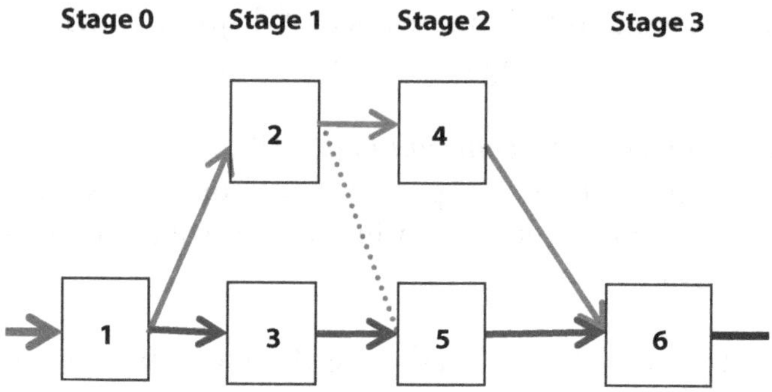

Figure 2.6 Network showing two disjoint paths and three redundant paths for S-D node pair (0, 6).

Disjoint paths: We get two disjoint paths (**1-2-4-6** and **1-3-5-6**) for source-destination node pair (1,6). These paths do not have any common switch other than source destination nodes.

Redundant and disjoint paths information give better knowledge about network capability to withstand minimum number of switch failures. Providing disjoint paths between any source and destination pair increases performance and reliability of the network. More disjoint paths enhance MIN fault tolerance as the common elements leading to all path failures are absent.

The disjoint paths are commonly used in two ways. The packets can be sent from the source to the destination nodes either using one of the disjoint paths or by using multiple packets simultaneously to all disjoint paths from the source to the destination nodes. In the former method if any switch/link fail they can tolerate faults by backtracking approach.

Let us explore the importance of disjoint paths and its importance in shuffle exchange network (SEN) [36, 71]. SEN as shown in Figure 2.7 is widely utilized class of MINs for practical purposes due to its simpler configuration. Shuffle exchange network (SEN) being a self-routing network, is based on two routing functions, shuffle and exchange. They provide only one path between any source-destination node pairs using 2×2 switches as basic elements.

SEN is a single-path MIN i.e. it has only one path between any pair of input and output nodes. SEN of size N×N consists of $\log_2 N$ stages, each stage containing N/2 SEs of size 2×2. This non fault

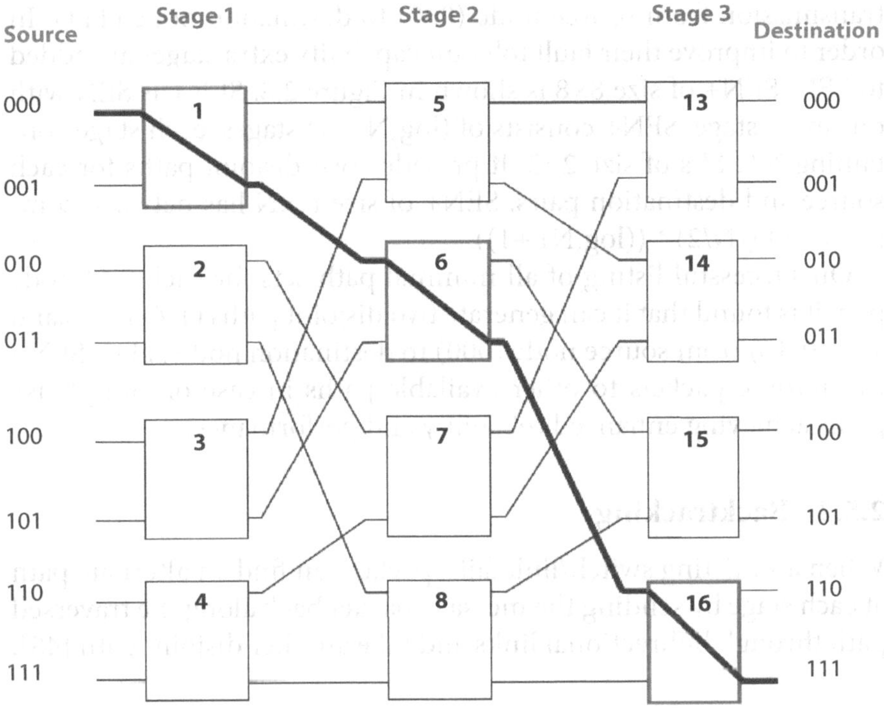

Figure 2.7 8×8 SEN showing single disjoint path for S-D node pair (0, 7).

tolerant MIN of size N×N has network complexity of $((N/2) * \log_2 N)$. This uses two operations using an m bit-wise address pattern of the inputs, $P_{m-1}P_{m-2}\ldots P_1P_0$, as follows:

$$S\,(P_{m-1}P_{m-2}\ldots P_1P_0) = P_{m-2}\ldots P_1P_0\,P_{m-1}$$

$$E\,(P_{m-1}P_{m-2}\ldots P_1P_0) = P_{m-1}P_{m-2}\ldots P_1P_0'$$

Here P_0' denotes complemented value. With shuffle (S) and exchange (E) operations, data packets are circulated until they reach destination. SEN can provide single path for any S-D node pair. Therefore, this network is poor fault tolerant it will result in loss of communication.

For example, let there exist a switch (i.e., switch 6) as busy/ faulty. Then, it will result in communication failure during message transmission from source node (000) to destination node (111). In order to improve their fault tolerant capability extra stages are added to SEN. SEN+ of size 8×8 is shown in Figure 2.8. SEN+ is SEN with one extra stage. SEN+ consists of ($\log_2 N + 1$) stages, each stage containing N/2 SEs of size 2×2. It provides two disjoint paths for each source and destination pairs. SEN+ of size N×N has network complexity of ((N/2) * (($\log_2$N) +1)).

On successful listing of all minimal path sets for each S-D node pair, it is found that it can generate two disjoint paths (1-6-12-16 and 1-5-10-16) from source node (000) to destination node (111). SEN+ can reroute packets to other available paths in case of faulty/busy paths achieving enhanced reliability and performance.

2.5.2 Backtracking

When an existing switch/link fails, packet can find an alternate path at each stage by sending the message packet back along the traversed path through bidirectional links and take another disjoint path [48].

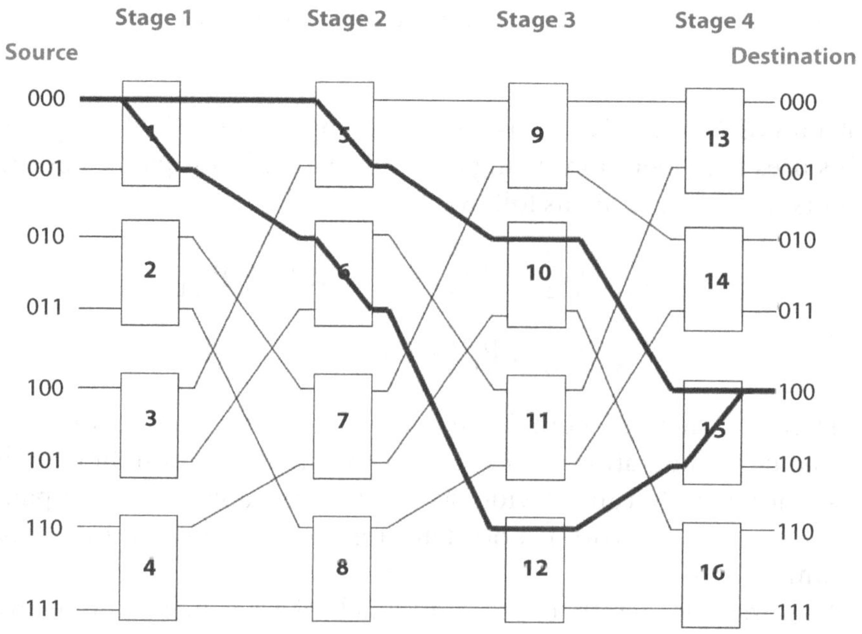

Figure 2.8 8×8 SEN+ showing two disjoint paths for S-D node pair (0, 4).

It degrades system performance on switch failures and increases rerouting hop count and collision rate.

2.5.3 Dynamic Rerouting

Although the method of sending packets concurrently increases reliability in disjoint paths network, the collision rate increases quickly as system loading increases. Dynamic rerouting is an alternative approach to design fault tolerant MIN under switch/link failures without backtracking. Dynamic rerouting method [50] corrects the routing tag in which the switch changes routing path to find an alternative path once a faulty or busy element is met. When a message packet encounters any switch/link faults, they automatically change their routing path and effectively handle collision problems.

2.6 MIN Topological Review on Disjoint Paths

Disjoint path information tells about number of minimum critical node failures required to cause network failure. An ***n-disjoint*** path fault-tolerant MIN has at least ***n*** number of such disjoint paths for any of its source destination node pairs. Therefore, it gives better information about network capability to withstand minimum number of switch failures.

The research efforts on developing multistage interconnection networks can be grouped into four categories based on disjoint paths:

1. Single-Disjoint Path MINs
2. Two-Disjoint Paths MINs
3. Three-Disjoint Paths MINs and
4. Four-Disjoint Paths MINs

2.6.1 Single-Disjoint Path Multistage Interconnection Networks

A wide number of single disjoint path MINs are available in literature. These networks provide only one disjoint path for at least one of input and output nodes pair. Banyan network [31], Baseline

[32], Delta networks [33], Cube network [34], Omega network [35], Shuffle-exchange network [36], Butterfly [37] and Dot network [38] are traditional self-routable MIN architectures. These basic networks provide only one path between any source-destination node pairs using 2×2 switches as basic elements. So, the network fails if any one switches fail, resulting in poor fault-tolerance.

Delta networks [33], is a subset class of Banyan interconnection networks, which allow a cost-effective direct connection between any processor and memory module. Figure 2.9 illustrates a b^nxb^n delta network made up of nb^{n-1}, b×b crossbar switches, arranged in *n* stages with b^{n-1} switches in a stage.

The link pattern used allows any source to be connected to any destination, where each b×b module connects an input to one of its b outputs depending on a single base-b digit taken from the destination address. It is a kind of blocking networks which has self- routing property. A Blocking network cannot realize all possible connections between inputs and outputs. This is because a connection between one free input to another free output is blocked by an existing connection in network.

Interconnection topology between every stage varies in every delta network. Following are popular types of topologically equivalent unidirectional Delta MINs:

- Omega (Figure 2.10) and their reverse networks (Flip)
- Baseline networks and Reverse Baseline networks
- Butterfly and Reverse Butterfly networks
- Cube and their reverse networks (Indirect Binary n-cube networks)

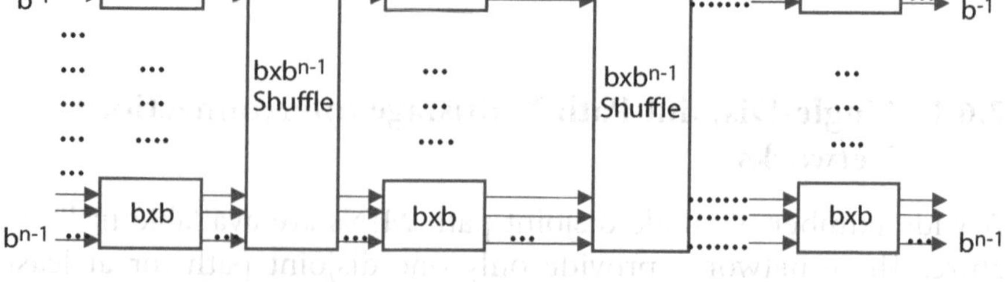

Figure 2.9 b^nxb^n Delta network.

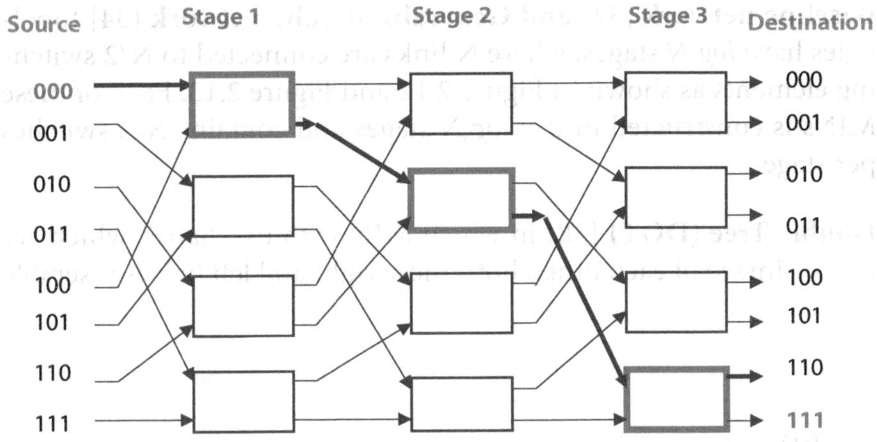

Figure 2.10 8×8 Omega network showing one disjoint path for S-D node pair (0, 7).

The Butterfly network [37] is a delta network, constructed with 2×2 crossbar connections, having $\log_2 N$ stages (Figure 2.11). Omega and Butterfly networks are isomorphic. This is a simple recursive MIN which results in a more economical implementation.

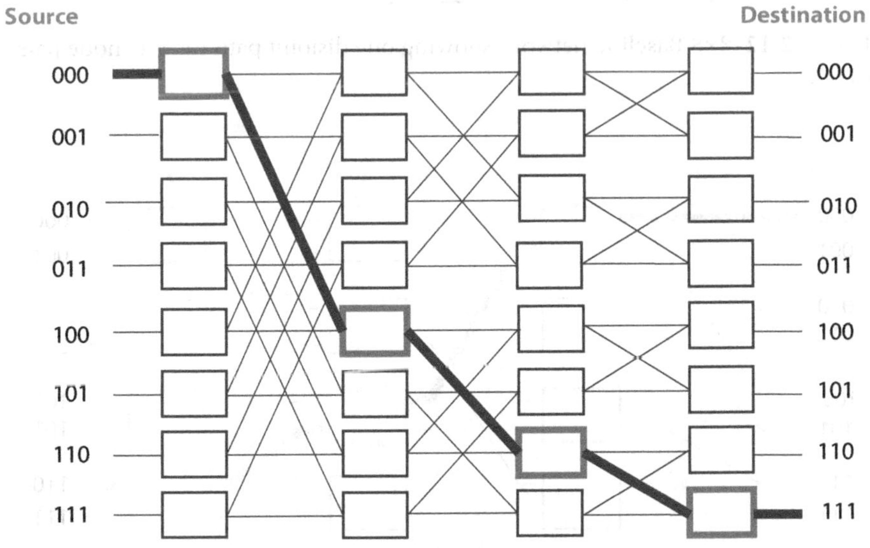

Figure 2.11 8×8 Butterfly network showing one disjoint path for S-D node pair (0, 7).

Baseline network [32] **and Generalized-cube network** [34] topologies have log_2N stages, where N links are connected to N/2 switching elements as shown in Figure 2.12 and Figure 2.13. Each of these MINs is constructed in n = log_2N stages and contains N/2 switches per stage.

Double Tree (DOT) [38] in which MIN with two halves which are mirror image of each other, both these right and left halves resemble

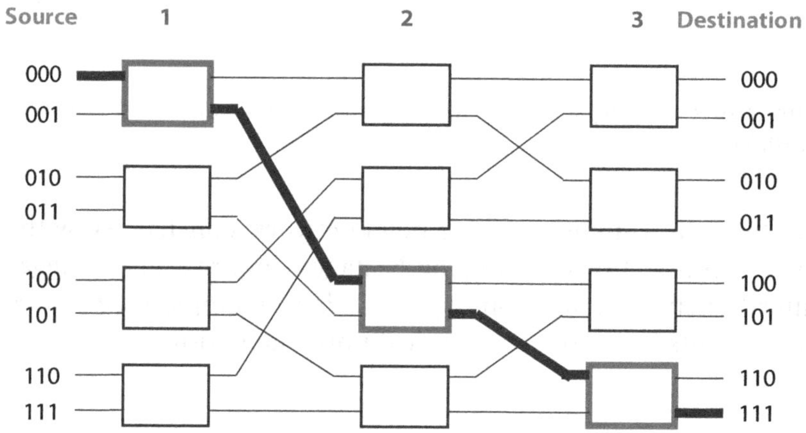

Figure 2.12 8×8 Baseline network showing one disjoint path for S-D node pair (0, 7).

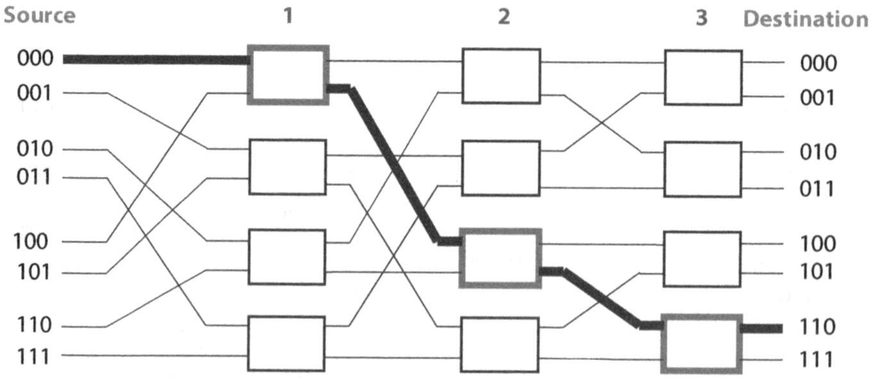

Figure 2.13 8×8 Generalized-cube network showing one disjoint path for S-D node pair (0, 7).

binary tree. Numbers of switches and links are $2^{n+1} - 3$ and $2^{n+2} + 2$ respectively. DOT Network is poor fault tolerant network providing single path for some source-destination pairs. Moreover, these networks have high probability of path contention and communication delay due to their longer path lengths as shown in Figure 2.14.

Theta network (THN) [39] consists of 2^{n-1}-1stages numbered from Stage 0 to Stage n, where $n = \log_2 N$. The switching elements used 2×2 size crossbars as shown in Figure 2.15. Thus, MIN with redundant

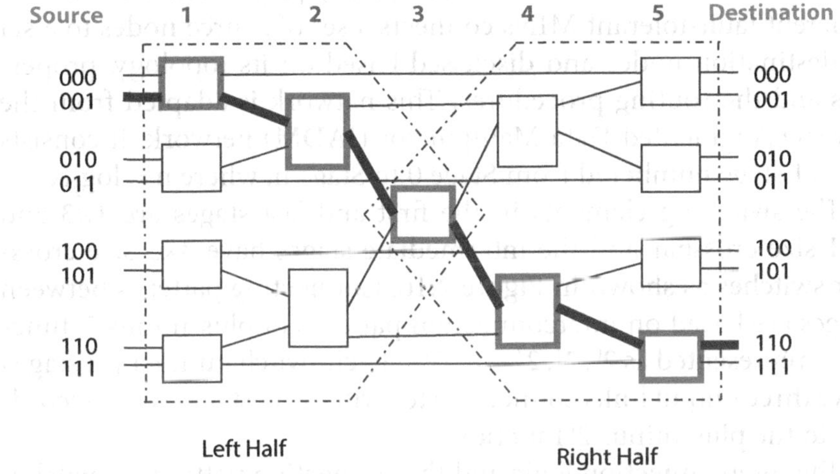

Figure 2.14 8×8 DOT network showing one disjoint path for S-D node pair (0, 7).

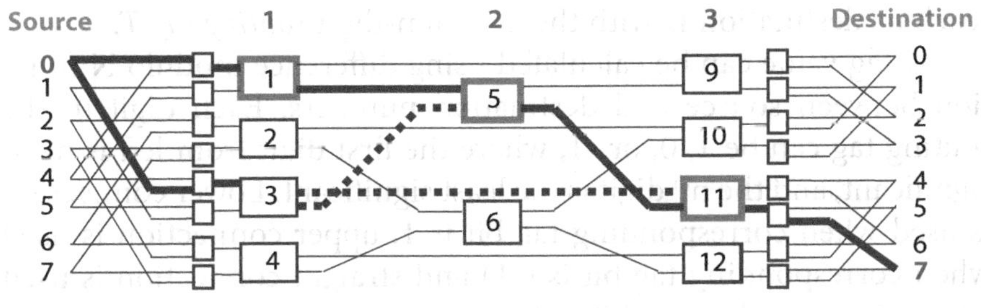

Figure 2.15 8×8 Theta network (THN) showing one disjoint path and three redundant paths for S-D node pair (0, 7).

paths is desirable since it can tolerate faults by routing requests over alternative paths enhancing reliability.

This network provides multiple paths by adding 2×1 size multiplexer at the initial stage and 1×2 size demultiplexers at the last stage. Number of multiplexers and demultiplexers used in this network is 2^n and their hardware complexity is more due to the addition of multiplexers and demultiplexers. Major advantage of this network is initial and final switch fault tolerance and provides full-access capability.

Gamma Interconnection network (GIN) [41] is another class of inherent fault-tolerant MINs connects a set of source nodes to a set of destination nodes and discussed based on its topology, properties and the routing procedures. This network is adapted from the Inverse Augmented Data Manipulator (IADM) network. It consists of n+1 stage numbered from Stage 0 to Stage n, where $n = \log_2 N$.

The switching elements in the first and last stages are 1×3 and 3×1 size crossbar and the intermediate stages have 3×3 size crossbar switches as shown in Figure 2.16. Connecting patterns between stages are based on interconnection patterns of plus-minus 2^i functions represented as $2^0, 2^1, 2^2.......2^{n-1}$. Each switch number j at stage i have three output links connecting to switches at stages (i+1) according to the plus-minus 2^i function.

The interconnection logic and the connection pattern of switches in all the stages are relevant and identical to the IADM [49, 72]. A full 3×3 crossbar element is used in GIN whereas each element used in IADM accepts only one input and directs it to one of the three outputs in the IADM network. In a GIN, message from source S can reach its destination D with the help of n-digit *routing tag T*.

The tag value can be calculated using difference modulo N function between source and destination numbers. Each digit in the routing tag can be 1, 0, or -1, where the first digit from left is most significant, and the nth digit is the least significant. Lower connection is used when corresponding tag bit is 1, upper connection is used when corresponding tag bit is (-1) and straight connection is used when corresponding tag bit is 0.

The relative position of nodes can be represented through *tag value* ($\boldsymbol{\delta}$), which is the difference between destination and source numbers

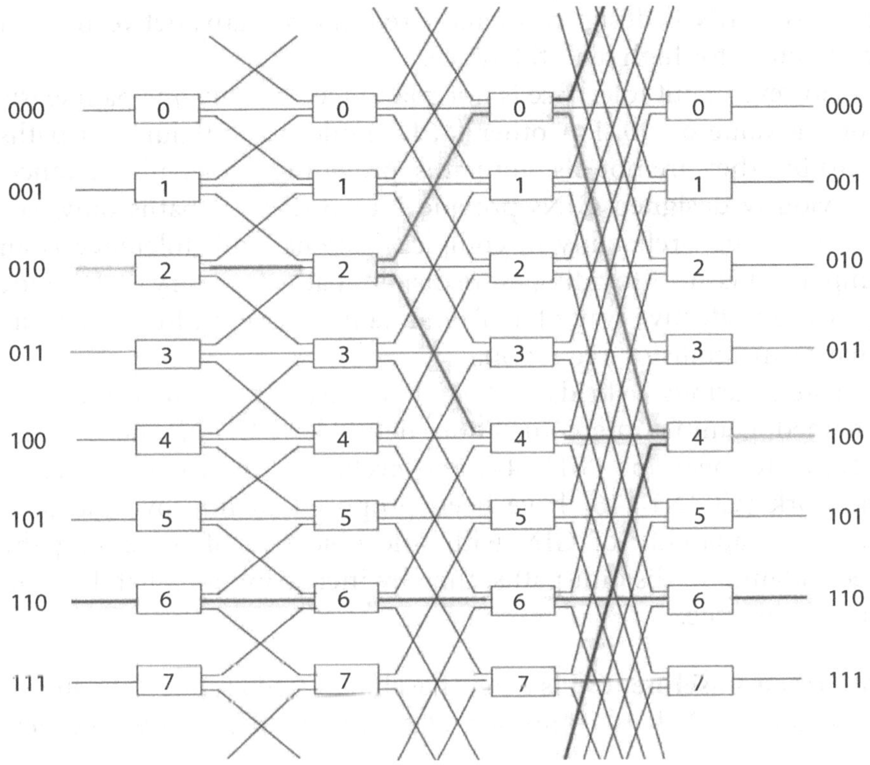

Figure 2.16 GIN showing paths for S-D node pairs (2, 4; δ = 2) and (6, 6; δ = 0).

δ=|D-S|. Now consider the Gamma network of Figure 2.14, for the flow of data from source node S = 2 to destination node D = 4 when N = 8. The difference between S and D is 2 and the binary form of 2 is 010. All redundant representations for the values 2 will be valid tags for routing data from source 2 to destination 4. The binary redundant forms for the tag value of 2 are (0,1,0), (0,-1,1) and (0,-1,-1). Therefore, for a tag value of 2, there are 3 redundant paths. However, GIN provides only one path when source and destination node sequence number are same, e.g. data flow from source node 6 to destination node 6 as shown in Figure 2.16.

In GIN, the symmetry of the network provides similar connectivity behavior for the source-destination (S, D) node pairs with equal relative distance between source and destination nodes. For example, connectivity behavior of (1, 2) node pair is same as (2, 3) node pair. The GIN is designed to allow redundant paths between input and output for high fault tolerance.

However, fault tolerance is not guaranteed as only one path exists for tag value δ = 0. For other (S, D) node pairs, redundant paths exist but they are not disjoint paths, providing poor fault tolerance. Previously designed GINs provide 2 or 3 disjoint paths only. For ensuring high reliability in complex systems, fault tolerance is an important issue. Providing more disjoint paths for every (S, D) node pair is an effective way of achieving fault-tolerance, higher performance and collision reduction.

Later, various hybrids of GIN like partially chained and fully chained gamma interconnection network (PCGIN and FCGIN [47]), Monogamma GIN [44] and cyclic gamma interconnection network (CGIN) [46] have been proposed to improve the fault tolerant capability of GIN with sole objective of increasing the redundant and disjoint paths, thereby increasing its reliability and fault-tolerance.

Modified Baseline [65] is a new baseline-based architecture shown in Figure 2.17. It is composed of symmetrical and asymmetrical

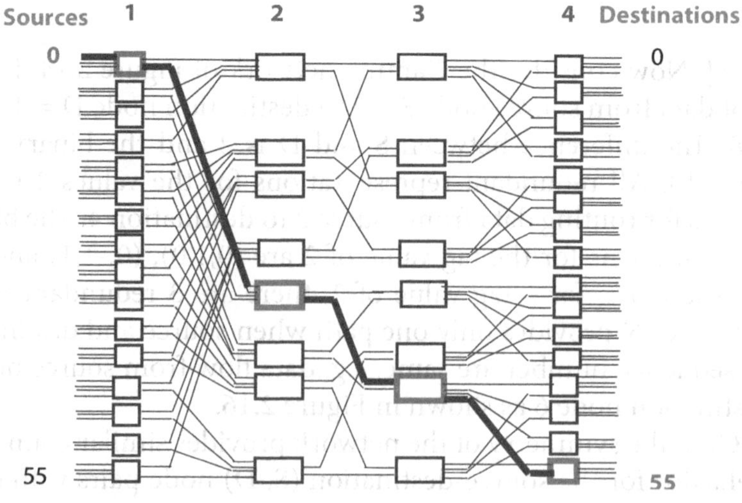

Figure 2.17 56×56 modified baseline showing one disjoint path for S-D node pair (0, 55).

switching elements proposed recently. Switching elements of various sizes are used. Baseline networks are generally blocking networks. But this network adapted from baseline network which is in strict-sense non-blocking network. But this network also provides only one path for every S-D pair.

Scalable crossbar network (SCN) [66] is a non-blocking MIN shown in Figure 2.18 proposed to solve the blocking and scalability issues. In this network 2×2 size crossbar switches are used as switching elements. This network also provides only one path for all source-destination (S-D) node pairs.

It is observed that most of the earlier designed MINs provided single disjoint path by using cheaper and simpler switches. Comparison of MINs is provided in Table 2.2 based on size of switches, number of switches and links used, and approximate hardware cost.

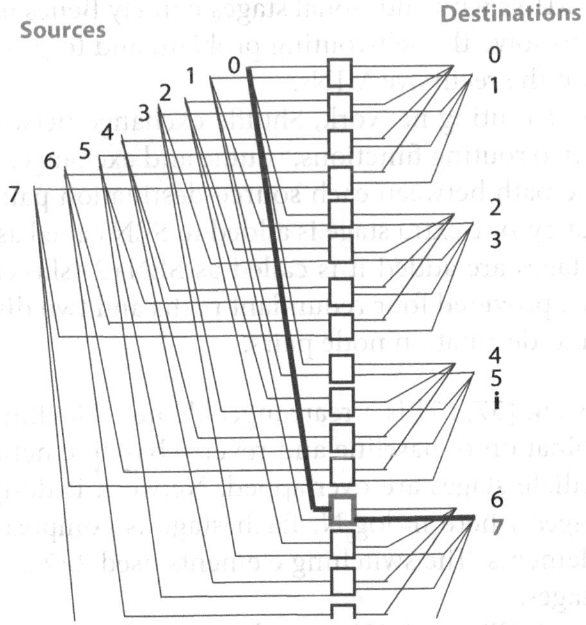

Figure 2.18 8×8 scalable crossbar network (SCN) showing one disjoint path for S-D node pair (0, 7).

2.6.2 Two-Disjoint Paths Multistage Interconnection Networks

A single disjoint path network may have one or more redundant paths but all paths have at least one critical switch in common whose failure leads to network failure resulting in poor fault tolerance. Providing more disjoint paths enhance MIN fault tolerance and reliability due to the existence of multiple disjoint paths between source–destination node pair and capability to withstand switch failures. These MINs have multiple redundant paths but they will possess at least two disjoint paths for any source–destination node pair. A wide number of two disjoint path multistage interconnection networks are available in literature.

Shuffle exchange network with one additional stage (SEN+) [71], extra stage cube (ESC) network with one additional stage [58] and Omega network with an additional stage (Omega+) provide two disjoint paths for any S-D node pairs. Limitation of the existing recursive MINs (including the baseline, banyan, and butterfly networks) is the self-routing and poor fault tolerance capability. Designing a new type of MINs with additional stages namely Benes network was introduced to solve the self-routing problem and to provide redundant paths on the recursive MINs.

Being a self-routing network, Shuffle exchange network [36, 71] is based on two routing functions, shuffle and exchange. But there is still a unique path between each source-destination pair. For creating redundancy, one extra stage is added to SEN called as SEN+ and when two stages are added it is called as SEN+2 (shown in Figure 2.19). SEN+2 provided four redundant paths and two disjoint paths for any source-destination node pairs.

Benes Network [57, 67] is a rearrangeable non-blocking MIN due to the combination of baseline and reverse baseline network where the intermediate stages are overlapped. Network is designed using (2n – 1) stages where $n=\log_2 N$. Each stage is composed of (N/2) switching elements. The switching elements used 2×2 size crossbars for all the stages.

This network (Figure 2.20) provides multiple paths at the initial stages but only one single path in final stages. From Figure 2.20, we can observe that there are four redundant paths but there are only

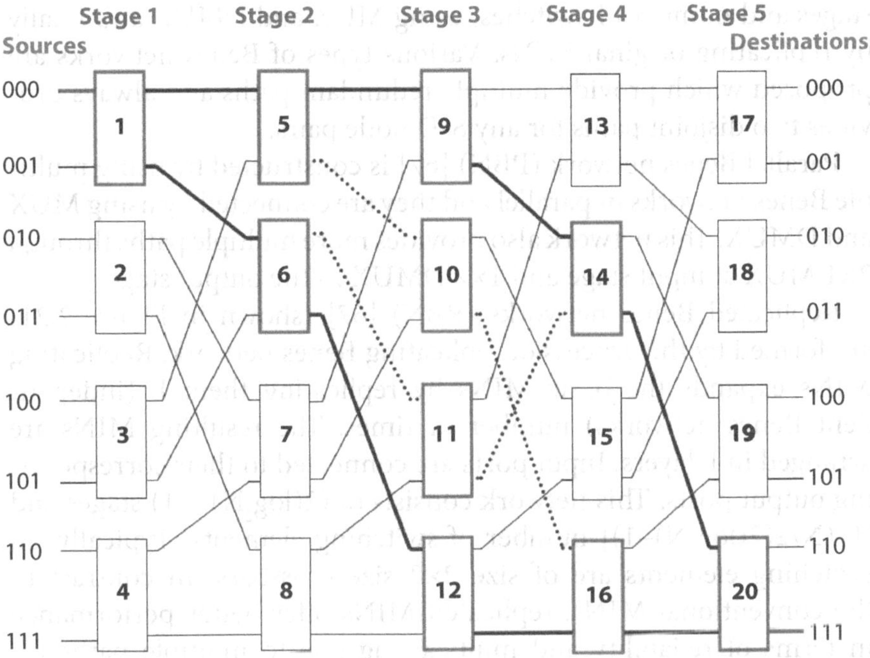

Figure 2.19 8×8 SEN +2 showing two disjoint paths and four redundant paths for S-D node pair (0, 7).

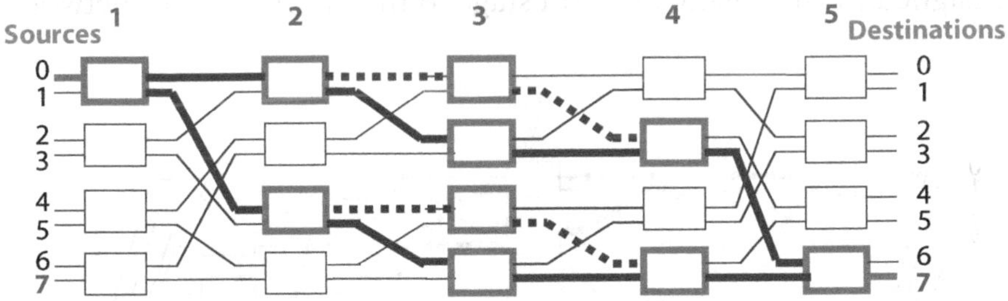

Figure 2.20 8×8 Benes network showing two disjoint paths and four redundant paths for S-D node pair (0, 7).

two disjoint paths. Benes network requires extra stages in its topology and hence its construction-cost is more expensive. These networks are not suitable for scalable systems. The larger network-sizes also lead to more delay-between the stages, and construction-cost.

To provide higher fault tolerance, many researchers later developed different fault tolerant Benes networks by increasing number of stages and number of switches, using MUX and DMUX and finally by replicating original MINs. Various types of Benes networks are proposed which provide multiple redundant paths and always provides two disjoint paths for any S-D node pairs.

Parallel Benes network (PBN) [67] is constructed by using multiple Benes networks in parallel and they are connected by using MUX and DMUX. This network also provides more multiple paths through 2×1 MUX at input stage and 1×2 DMUX in the output stage.

Replicated Benes networks (RBN) [67] (shown in Figure 2.21) are formed by the process of replicating Benes network. Replicating MINs expands the basic MINs by replicating them L (independent Benes networks) number of times. The resulting MINs are arranged in L layers. Input ports are connected to their corresponding output ports. This network consists of $(2(\log_2 N) - 1)$ stages and $[L\ (N/2(2(\log_2 N)-1)]$ number of switching elements. Typically, all switching elements are of size 2×2 size crossbars. In contrast to the conventional MINs, replicated MINs offer better performance in terms of reliability and multicasting due to multiple paths for source-destination node pairs. However, there is major flaw in replicated MINs that they need more switching power in the last stages.

Recently, Extra-stage Benes network (EBN) [67] (Figure 2.22) designed by increasing an extra stage to the original Benes network

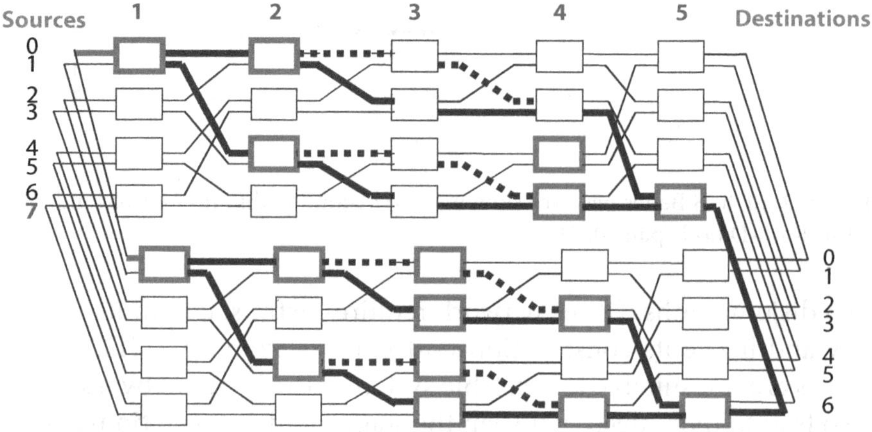

Figure 2.21 8×8 RBN showing two disjoint paths and eight redundant paths for S-D node pair (0, 7).

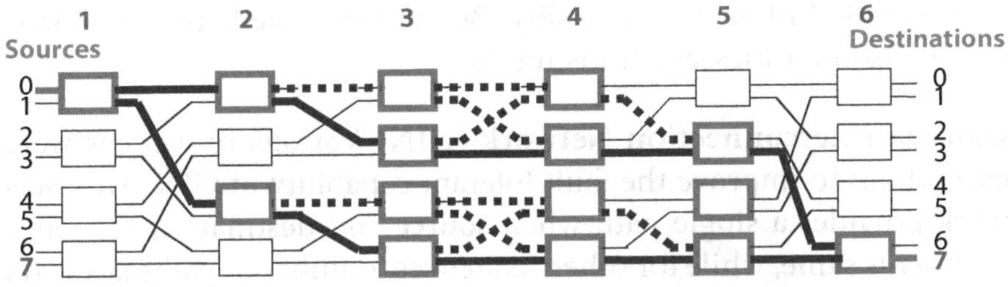

Figure 2.22 8×8 EBN showing two disjoint paths and eight redundant paths for S-D node pair (0, 7).

to provide multiple paths between each source–destination pairs. Space complexity is the major drawback of this network.

Phi network (PHN) [40] consists of $\log_2 N$ stages. The switching elements use 2×2 size crossbars in first and last stages and 3×3 size crossbars in rest of the stages. This network provides five redundant paths and two disjoint paths for any source-destination node pairs. They also used 2×1 size multiplexer at the initial stage and 1×2 size demultiplexers for routing requests over alternative paths. Number of multiplexers and demultiplexers used in this network is 2^n and their hardware complexity is more due to the addition of multiplexers and demultiplexers.

Major advantage of this network (Figure 2.23) is through initial and final stage switch fault tolerance. Multiplexers and Demultiplexers

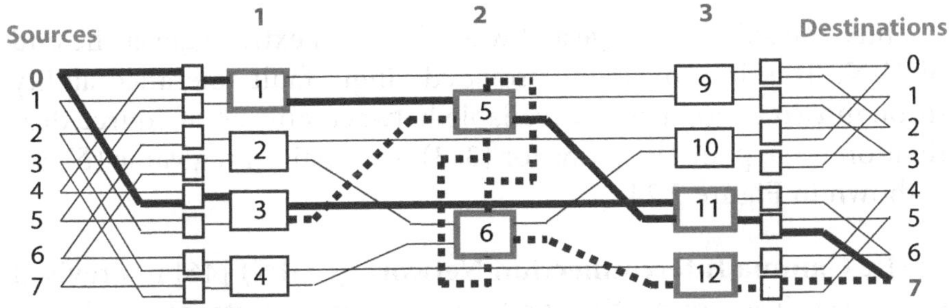

Figure 2.23 8×8 Phi network showing two disjoint paths and five redundant paths for S-D node pair (0, 7).

also provide full-access capability. But any intermediate switch failure ceases their message transmission.

Gamma Interconnection Network (GIN) Various researches were carried out to improve the fault tolerant capability of GIN. This network provides a single path when source and destination sequence number is same, while for other sequences, number of paths is always more than one. Various hybrids of GIN have been proposed with hardware cost almost same as GIN are analyzed here. Family of GINs provided two disjoint paths includes extra stage gamma interconnection network (ESGIN) [42], B-network [43], partially chained gamma interconnection network (PCGIN) [47], fully chained gamma interconnection network (FCGIN) [47], balanced gamma network (BGIN) [42], cyclic gamma interconnection network (CGIN) [46], no-backtracking GIN (NBGIN) [48], minimal links traversed dynamic rerouting GIN (MINGIN) [50], Incomplete GIN [55] and combining switches Multistage interconnection network (CSMIN) [51]. These networks provide multiple redundant paths and two disjoint paths always with capability to withstand switch failures.

Extra-Stage Gamma Interconnection Network (ESGIN) [42] was constructed to overcome two important problems of GIN when compared to Gamma network [41]

- There exists only one path when $\delta = 0$,
- Number of paths for even tag values is less compared to the odd tag values.

Thus, the solution suggested was adding an extra stage at the end of GIN. This improvement provided single fault tolerance ability through two disjoint paths available between any set of source destination node pair. The paths for (3, 3) node pair of tag value (δ) = 0 is shown in Figure 2.24.

Cyclic Gamma Interconnection Network (CGIN) [46] is a revised GIN [41]. The connecting pattern between the stages exhibits a cyclic feature to tolerate any single switch fault. It is designed to provide multiple disjoint paths without increasing hardware cost. The

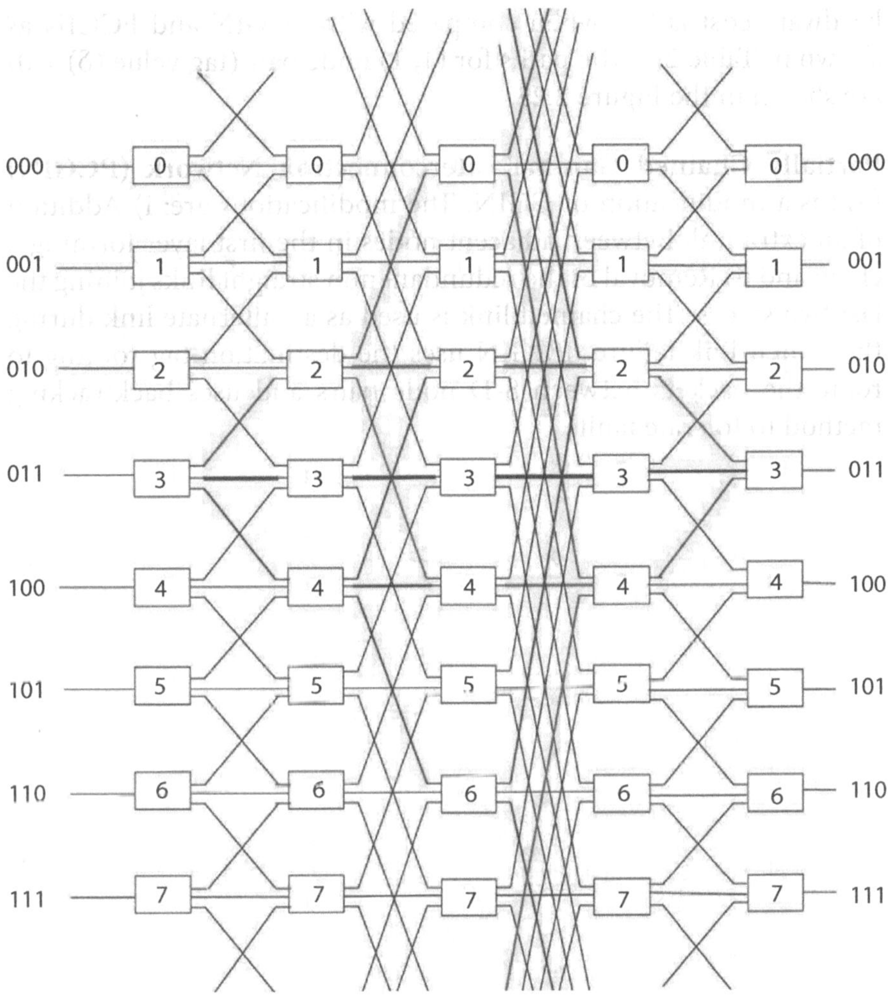

Figure 2.24 Extra-stage GIN showing nine paths for tag value (δ) = 0.

repeated interconnection pattern between any two stages provides at least 2 disjoint paths between any pair of source and destination nodes. It uses the destination tag routing and re-routing (routes a request under the guidance of the destination address only).

But the routing algorithm is complicated and there is no strong reroutability. It uses the backtracking scheme to find an alternate path whenever a fault is encountered in the existing path. The pin count reduces due to the repeated interconnection pattern and the hardware cost is less when compared with PCGIN and FCGIN as shown in Table 2.2. The paths for (1, 1) node pair (tag value (δ) = 0) are shown in the Figure 2.25.

Partially Chained Gamma Interconnection Network (PCGIN) [47] is a modification of CGIN. The modifications are: i) Addition of an extra link between adjacent nodes in the first layer forming a chain and ii) Removal of the redundant non straight links joining the last two stages. The chained link is used as an alternate link during the switch/link failure. PCGIN uses the destination tag routing to route the packets between S-D node pairs and uses backtracking method to tolerate faults.

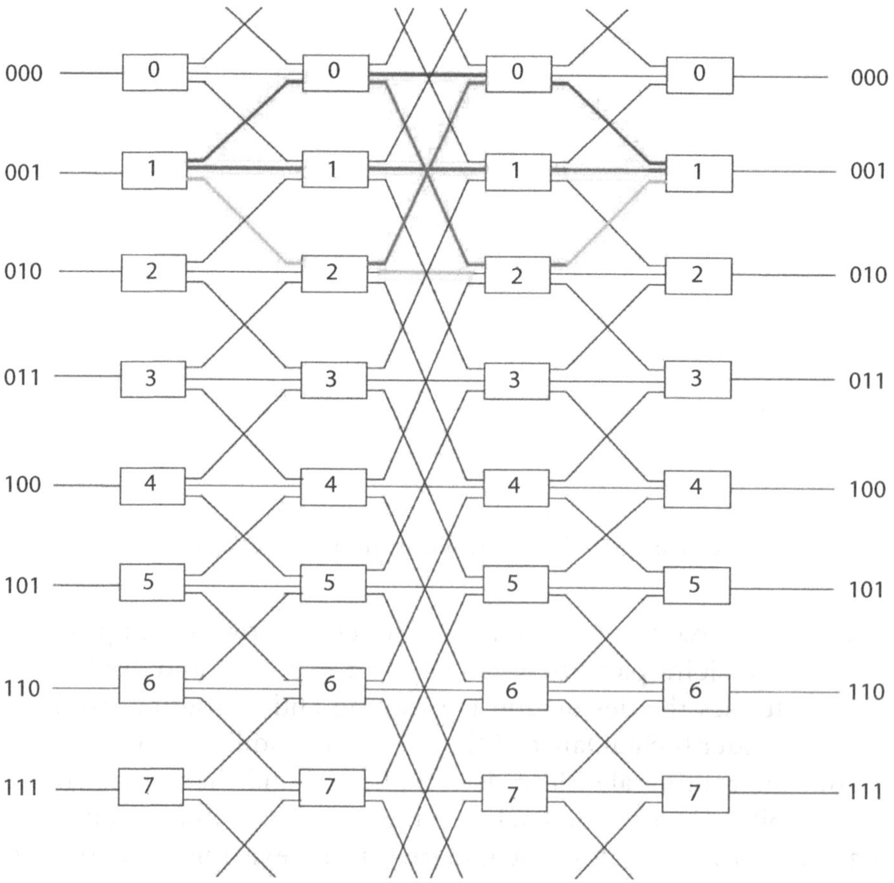

Figure 2.25 CGIN showing five paths for S-D node pair (1, 1; δ = 0).

In the backtracking method, a packet goes backward to the previous stage to find an alternate path, when the existing path is found faulty. The backtracking method degrades system performance. PCGIN can tolerate at least one link/switch fault as it provides at least two disjoint paths between any source and destination node pair. The two disjoint paths will not intersect at the intermediate switches before they reach their destination is shown in Figure 2.26.

Fully Chained Gamma Interconnection Network (FCGIN) [47] as shown in Figure 2.27 uses a fully chained approach. It connects back one of the non-straight links to the same stage (not connecting to next stage) in a chain formation similar to the first stage of PCGIN. Due to chaining at every stage, it provides distributed control and dynamic rerouting by avoiding backtracking penalties. FCGIN has many advantages over CGIN and PCGIN such as lower hardware

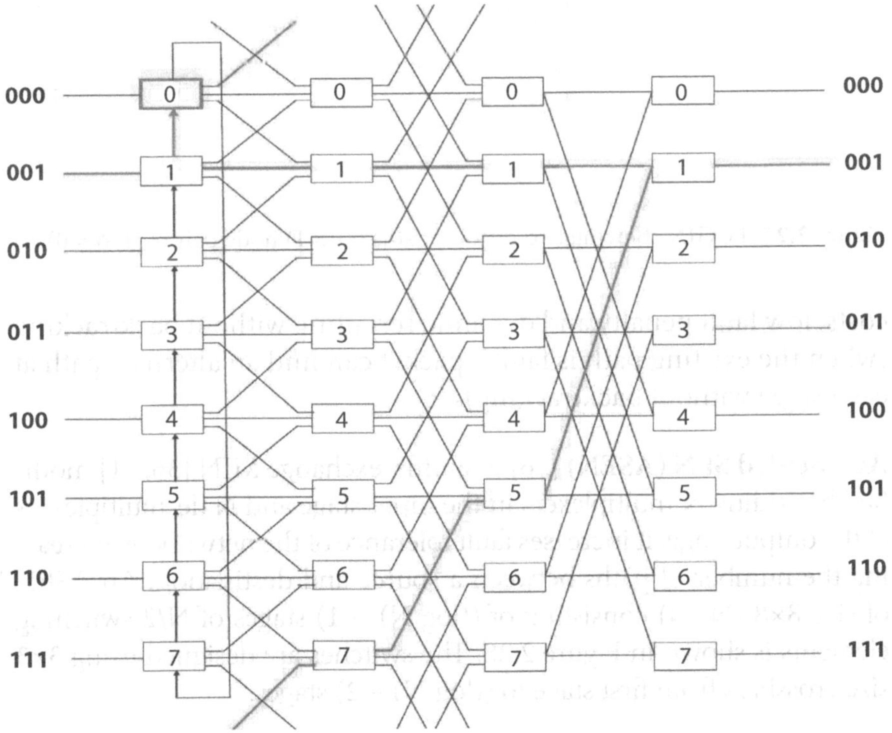

Figure 2.26 PCGIN showing two disjoint paths for S-D node pair (1, 1; δ = 0).

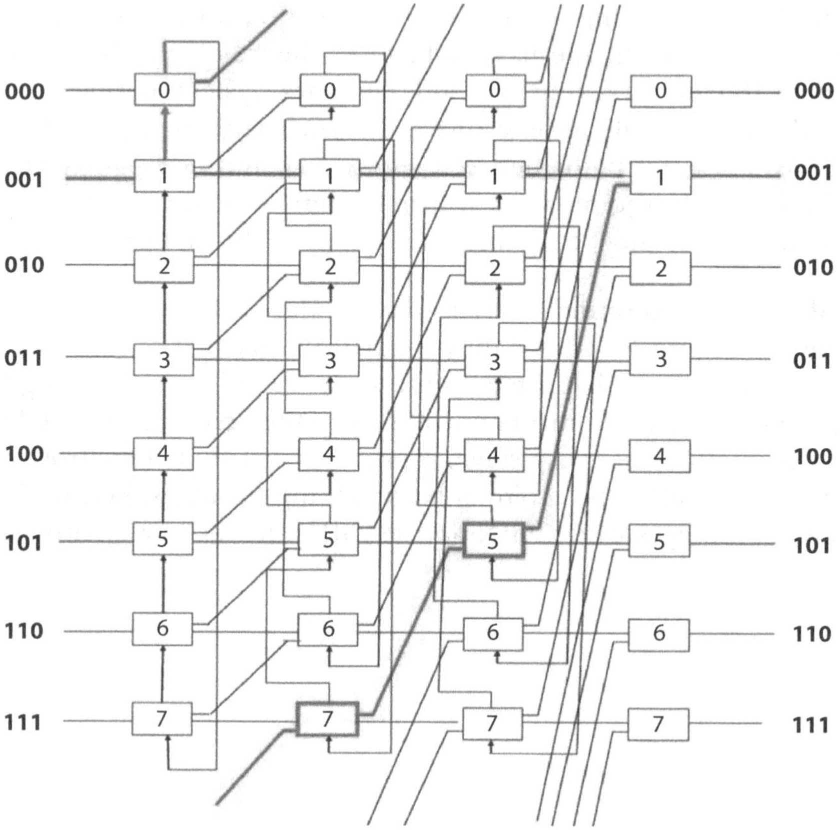

Figure 2.27 FCGIN showing two disjoint paths for S-D node pair (1, 1; $\delta = 0$).

costs, low fault penalty and dynamic rerouting without backtracking (when the existing path is faulty, packet can find an alternate path at each stage without backtracking).

Augmented SEN (ASEN) [36] is shuffle-exchange MIN [36, 71] modified by adding N multiplexers in the input stage and N de-multiplexers at the output stage. It increases fault tolerance of the network by increasing the number of paths between a source and destination. An ASEN of size 8×8 (N×N) consisting of $((\log_2 N) - 1)$ stages of N/2 switching elements is shown in Figure 2.28. The switches are designed using 3×3 size crossbars from first stage to $((\log_2 N) - 2)$ stages.

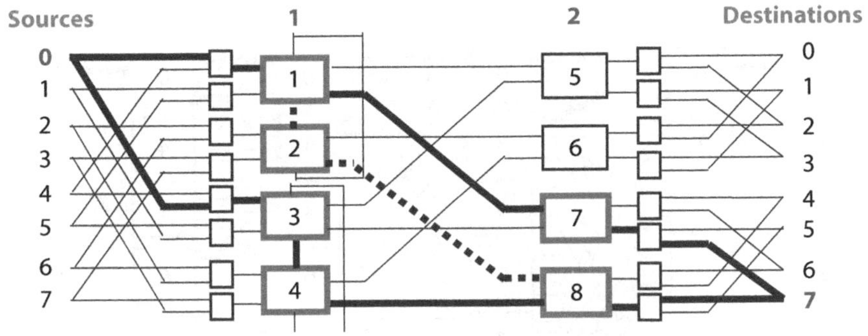

Figure 2.28 8×8 ASEN showing two disjoint paths and four redundant paths for S-D node pair (0, 7).

Its final stage uses 2×2 size crossbars. It uses 2×1 multiplexer at the initial stage and 1×2 demultiplexers at the $((\log_2 N) - 1)$ stage providing multiple paths between all source and destination pairs. However, addition of MUX and DMUX is disadvantageous as it increases hardware complexity and communication delay.

Kappa Network [56] is another popular switching topology which tolerate arbitrary single switch faults. This network is a modified gamma interconnection network (GIN) [41] of n+1 stages as shown in Figure 2.29. The switching elements in the first and last stages are 1×4 and 4×1 size crossbars while intermediate stages use 4×4 size crossbar switches. Kappa network is designed to provide symmetric redundant paths which overcome the problem of asymmetric redundant paths availability of GIN for different S-D node pairs. However, its hardware complexity is high compared to other MINs.

Tandem Omega [62] in which the first network in tandem can be the omega network appended with the banyan exchange while the second is the reversed omega network. Tandem connection between $2^n \times 2^n$ omega network [35] and $2^n \times 2^n$ reverse omega network [32] adjoined with varying $2^n \times 2^n$ banyan exchange network connections is shown in Figure 2.30. However, they provide only two disjoint paths between any S-D node pairs with their high extent network.

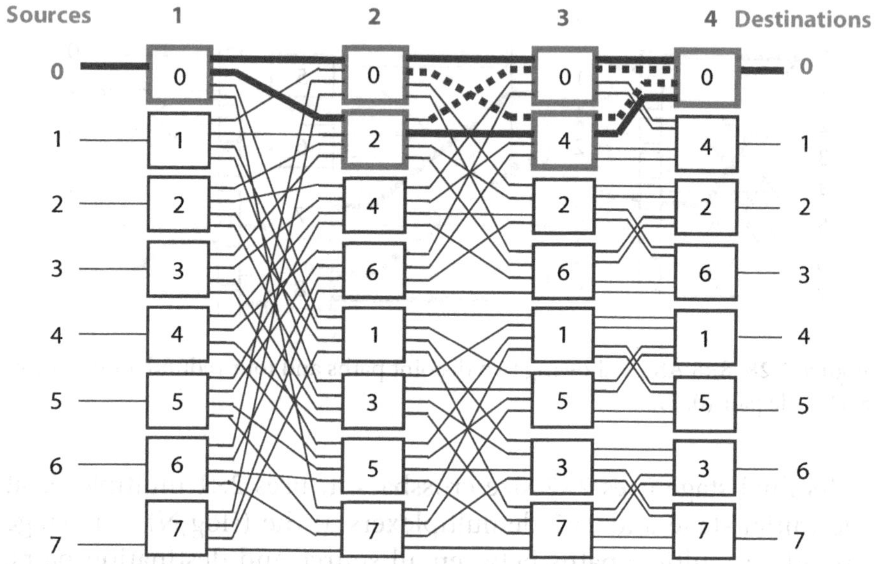

Figure 2.29 8×8 Kappa showing two disjoint paths and eight redundant paths for S-D node pair (0, 0).

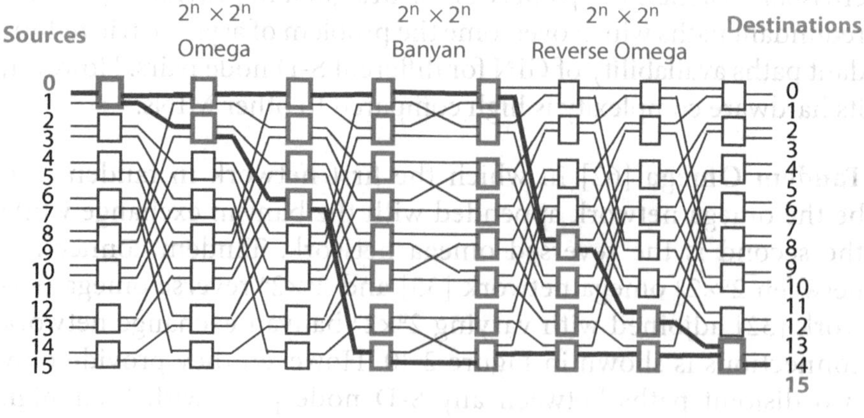

Figure 2.30 16×16 Tandem Omega showing two disjoint paths for S-D node pair (0, 15).

2.6.3 Three-Disjoint Paths Multistage Interconnection Networks

This family of MINs always possesses minimum three disjoint paths for any source–destination node pairs. Three disjoint paths between any input/output pair, can tolerate minimum two switch faults. Some notable available MINs from literature are discussed in this section. Enhanced IADM [49, 72] is adapted from the Inverse Augmented Data Manipulator (IADM) network. It consists of n + 1 stage numbered from Stage 0 to Stage n, where $n = \log_2 N$. The switching elements in the first and last stages are 1×3 and 4×1 size crossbar and the intermediate stages have 3×5 and 5×4 size crossbars as shown in Figure 2.31.

GIN [41] and IADM [72] are different in the sense that GIN uses a full 3×3 crossbar, whereas in the IADM uses 1x3 crossbars. The connection pattern in both networks is identical. The IADM network using 5×4 crossbar switches in intermediate stage guarantees better fault tolerance, but the cost became prohibitive.

Three-Disjoint Gamma Interconnection Networks (3DGIN) [49] is a modified GIN [31] that provides disjoint paths to tolerate minimum two switches or link faults between S-D node pairs. The 3DGIN is similar to GIN except its first stage uses 2×4 switches (by converting two switches into one at the first stage) instead of 1×3 switches used in GIN. 3DGIN also provides 3-disjoint paths and double-fault tolerance capability. It has lesser hardware cost compared to GIN. It uses backtracking method on occurrence of any switch fault to find a new path with help of rerouting tag.

Three-Disjoint Cyclic Gamma Interconnection Network (3DCGIN) [56] is modified CGIN, as shown in Figure 2.32. It provides an alternate source at the first stage to tolerate faults. In the first stage, the first node is connected with the fifth node with an alternate link providing redundant paths and re-routing facility for any S-D node pair. The connecting patterns between different stages are same as CGIN. To avoid the transmission delay and initial switch fault in CGIN, the alternate source can be used in 3DCGIN. Thus, there is no interruption of transmission on occurrence of switch faults in the first stage

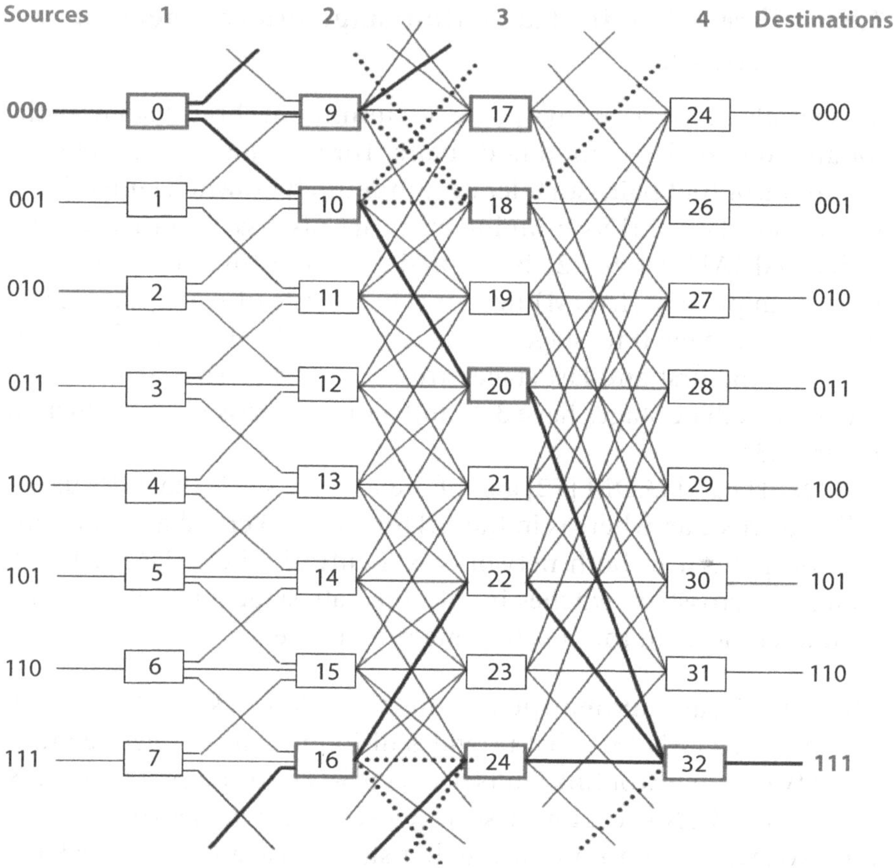

Figure 2.31 8×8 enhanced IADM network showing three disjoint paths and eight redundant paths for S-D node pair (0, 7).

as well. 3DCGIN is compatible with both distance tag routing and destination tag routing techniques.

Accurate routing and path congestion reduction is achieved by dynamic re-routing and pre-computing routing tags. This network provides at least 3 disjoint paths and guarantees at least 2 switches/link fault tolerance acting as a double fault tolerant network. However, it has higher hardware cost compared to other 3DGINs.

Two-Dimensional Ring-Banyan Network [61] is a fault-tolerant switching network provides multiple redundant paths using chained

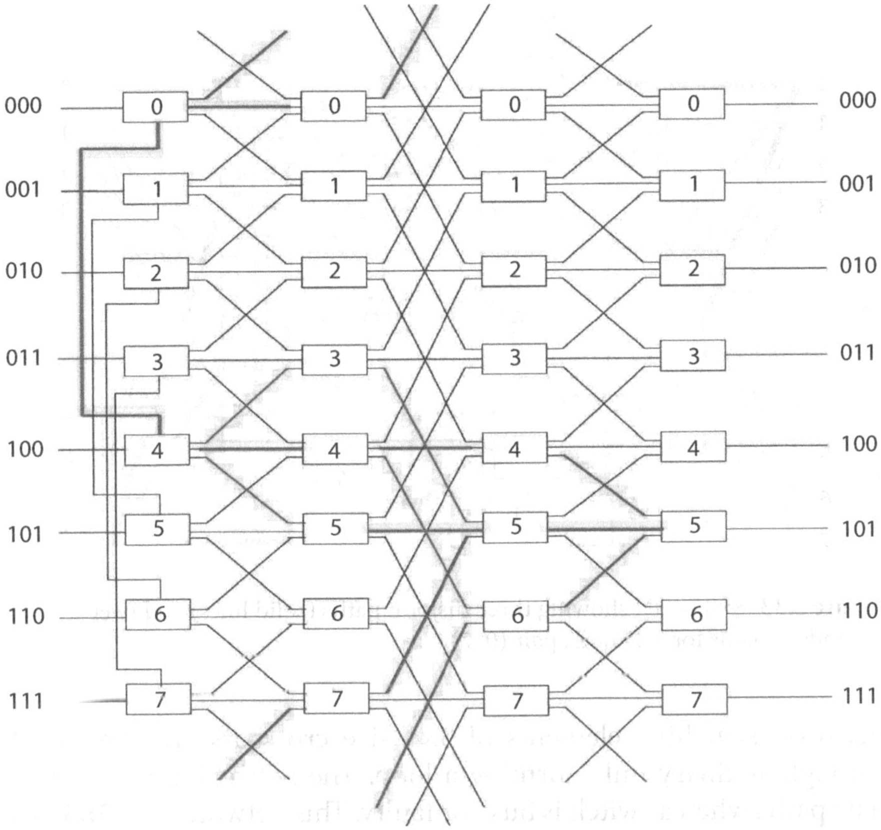

Figure 2.32 3D-CGIN showing three disjoint paths (red), three alternate paths (green) for S-D node pair (0, 5; δ = 5).

link connections between the switches of all the stages. However, they provide only three disjoint paths for any S-D node pair. This completely adaptive self-routing network is very complex due to fully chained link topology at all stages. All stages use 4×4 size crossbars which provide redundant paths using chained pattern to provide better performance and higher reliability.

Improved Extra Group Network (IEGN) [68] is a new dynamically re-routable regular MIN (Figure 2.33) providing better fault tolerance by use of auxiliary links in both the initial and final stages of the network. The approach of utilizing extra links at final stage makes perfect connection to the penultimate stage switches. In this

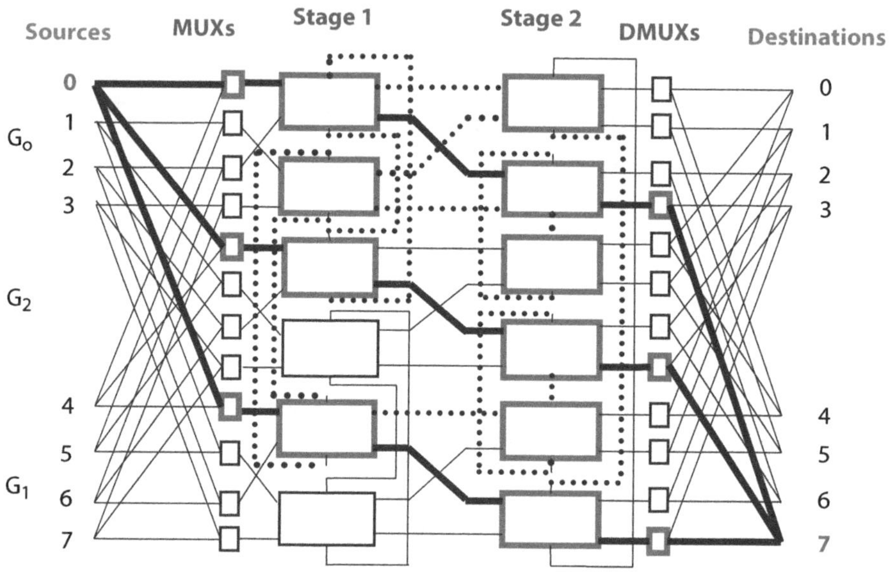

Figure 2.33 8×8 IEGN showing three disjoint paths (solid lines) and twelve redundant paths for S-D node pair (0, 7).

network, switching elements of 3×3 size crossbars are connected through auxiliary links forming a loop. The network utilizes alternate paths when a switch is busy or faulty. This network uses 3×3 SEs and consists of $((\log_2(N/m))$ stages. It consists of multiplexers and de-multiplexers in input and output stages respectively.

Pars Network [69], recently introduced regular fault tolerant self-routing MIN which surpassed IEGN in terms of cost, fault-tolerance and terminal reliability. A pars network (Figure 2.34) consists of $((\log_2 N)\text{-}1)$ stages of (3N/4) switches. The switching elements used 2×2 size crossbars. They also consist of (3N/2) multiplexers and de-multiplexers in input and output stages respectively. For larger network-sizes the hardware complexity is more due to addition of MUX and DMUX.

For generalized baseline network [32] topology, there is only one path from any input to any output. Therefore, this network is less fault tolerant network and the input and output will be unable to communicate during switch failure. As the solution, **Augmented Baseline**

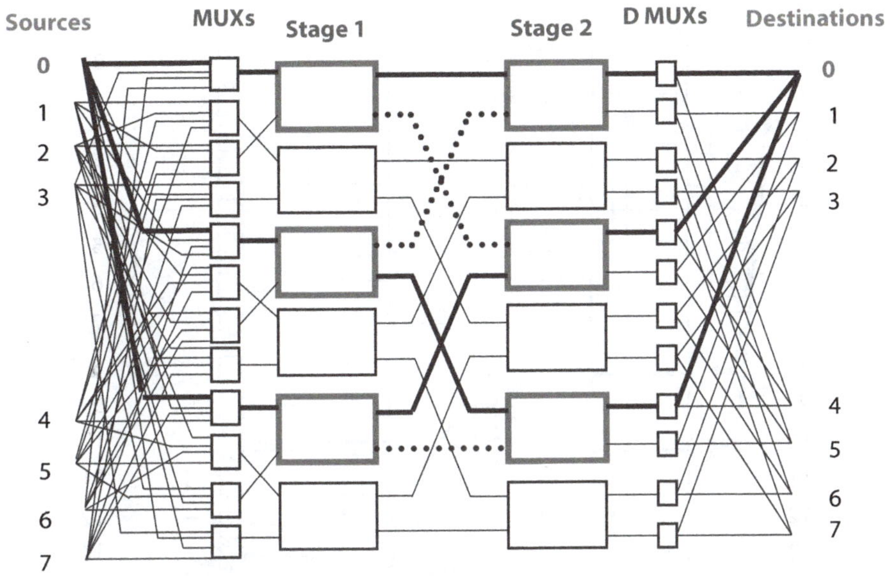

Figure 2.34 8×8 Pars network showing three disjoint paths (solid lines) and six redundant paths for S-D node pair (0, 0).

Network (ABN) [59] network was introduced, which is a fault-tolerant network derived from the generalized-baseline network. ABN is another irregular network which uses 3×3 and 2×2 size crossbars at subsequent stages. ABN of size 16×16 network is shown in Figure 2.35. This network includes MUX of size 4×1 at the initial stage and DMUX of size 1×2at the final stage which increases the network fault tolerance by increasing the number of paths between a source and destination. Average path length of ABN is greater than baseline network and number of successful connections between source and destination is high in case of ABN.

2.6.4 Four-Disjoint Paths Multistage Interconnection Networks

Fault tolerance can be achieved by creating redundancy in system components so that if a component fails, the successor components can be used. According to these discussions, we deduce MINs require

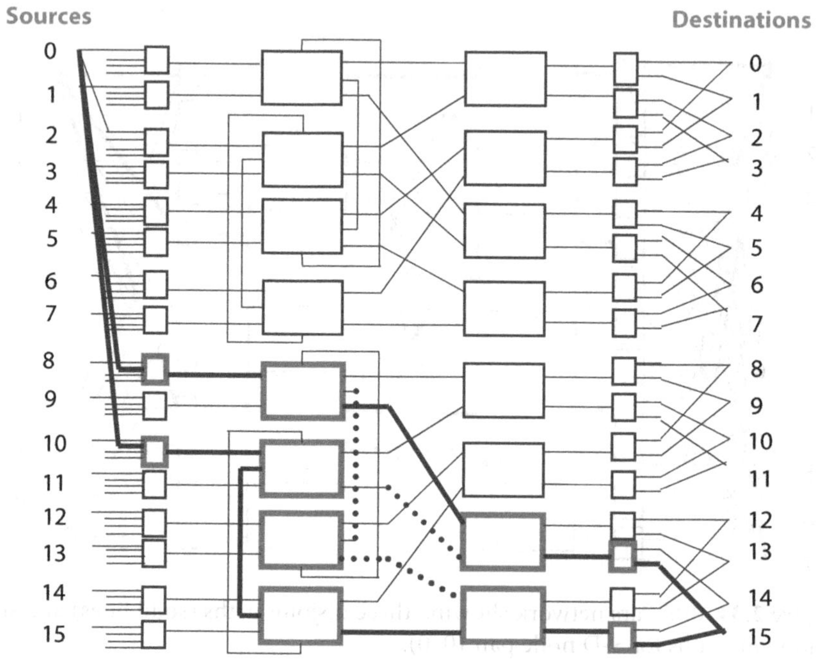

Figure 2.35 16×16 ABN showing two disjoint paths and four redundant paths for S-D node pair (0, 15).

three important factors to be used in modern systems: (1) High reliability; (2) Efficient multicasting; and (3) Cost-effectiveness.

The multi path networks are fault tolerant and more reliable because they provide more redundant and disjoint paths. This has attracted focus of the researchers. Compared to conventional MINs, 4-disjoint path MINs offer better performance in terms of reliability and multicasting as multiple paths exist for a source-destination pair.

Extended Generalized Shuffle (EGS) [63] is a class of strictly non-blocking multistage interconnection networks using optical switching architecture. This network is designed using symmetrical switching elements interconnected by inter-stage links. They used 2×2 size crossbars in all the stages. These crossbars are mutually interconnected through EGS inter-stage pattern. However, this network uses external splitters and combiners of same size to connect inputs and outputs. Thus, 8×8 EGS network has 4 stages using 1×4 splitters and 4×1 combiner, and 16 switching elements of size 2×2 size crossbars as shown in Figure 2.36. It provides multiple redundant paths and four disjoint paths for all input-outputs node pairs. However, inclusion of splitters, combiners and high number of switching elements increased hardware complexity and space.

Charles Clos Network [73] is a non-blocking array of cross-points as shown in Figure 2.37. It is used in commercial telephone systems.

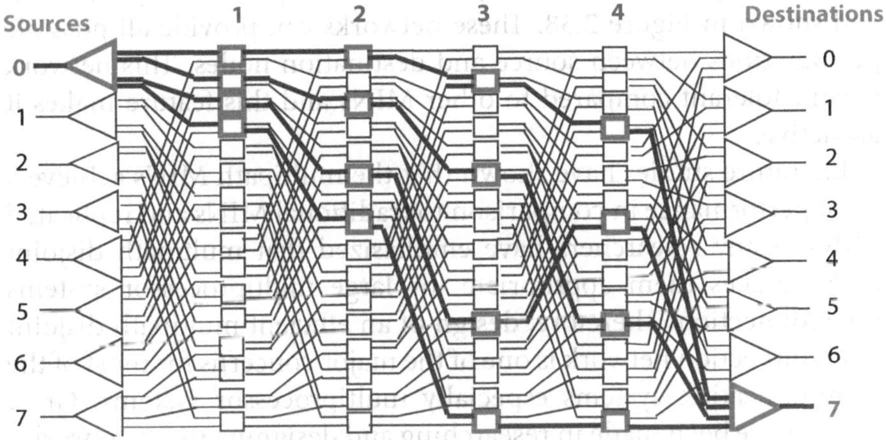

Figure 2.36 56×56 EGS Network showing four disjoint paths for S-D node pair (0, 7); Disjoint paths and three redundant paths for S-D node pair (0, 7).

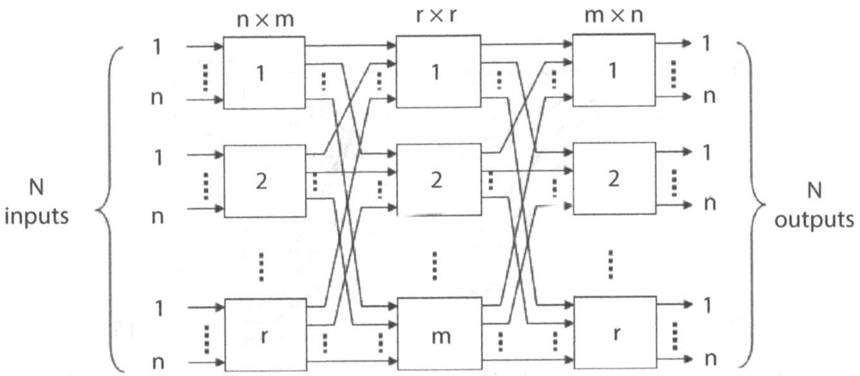

Figure 2.37 Three stage clos network. Disjoint paths and three redundant paths for S-D node pair (0, 7).

Clos networks are defined by three stages (n,*m*,*r*) and represented with integer's *n*, *m*, and *r*. Here *n* denotes the number of inputs in each of *r* first-stage switches. Number of second-stage switches is represented by *m*. Clos network provides multiple paths with a high design layout.

Clos networks have two main problems that prevent them from being the practical solution in this domain. Generally, Clos network is more expensive, more complex and difficult to control than rearrangeable non-blocking and blocking MINs. Recently new type of non-blocking **Clos network** [64] is developed represented by C (4, 4, 4) and shown in Figure 2.38. These networks can provide all possible permutations between source and destination nodes. This network is fault-tolerant compared to other MINs and this feature makes it attractive.

Literature studies have shown that the multipath MINs achieve a good performance in comparison to traditional MINs and replicated MINs. Many researchers have emphasized that multipath disjoint MINs makes them appropriate for large multiprocessor systems interconnection. Therefore, design of an efficient multipath disjoint interconnection network is one of the major concerns for most of the communication systems especially multiprocessor systems. Great efforts have been made in researching and designing these networks. However, cost of network is major constraint when designing multipath multi-fault tolerant MINs.

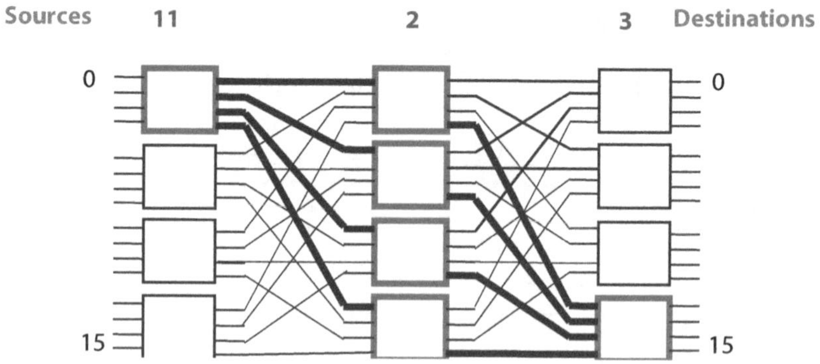

Figure 2.38 16×16 clos network showing four disjoint and redundant paths for S-D node pair (0, 15).

2.7 Hardware Cost Analysis

In the choice of an interconnection network for use in large-scale systems, balance need to achieved between cost and performance. Cost plays a predominant parameter in the choice of a large-scale interconnection network layout design. High fault tolerance and reliability coming at the expense of high cost may not be affordable in practice. The modest cost of the MINs makes them appropriate for large multiprocessor systems.

A common method [74] to estimate the cost of a MIN is to calculate the switch complexity with an assumption that the cost of a switch is proportional to the number of gates involved, which is roughly proportional to the number of cross-points within a switch. Cost of a network is calculated using design details such as number of switches, links and connection points.

Total cost of switches is calculated for all the MINs as summation of product of number of input pins (I) and output pins (O) of all switches (S) used in the network. For example, a 2×2 switch has four units of hardware cost, whereas a 2×3 switch has six cost units. A MUX and DMUX of size m×1 and 1×m corresponds to m units of cost. The crossbar switch sizes and the number of crossbars switches of various MIN architectures are provided in Table 2.2. A sample cost calculation Clos network [64] in Figure 2.36 is shown below.

Total no of 4×4 switching elements = 12

Cost = I×O×S = 12×4×4 = **192** units.

Total cost of Clos network [64] is **192** units.

From Table 2.2 it is noted that the hardware complexity of multiple-path MINs is more than unique-path MINs. Hardware complexity of the design mainly depends on the number and sizes of switches used, multiplexers (Mux) and de-multiplexers (Demux).

Mux-Demux: Among multiple-path MINs [36, 59, 63, 67–69], some used Mux at the initial stage and Demux at the last stage which increased the cost.

Table 2.2 Multistage interconnection networks topology comparison.

Network types	**Size**	**Switch size**	**Switch/Links/ Mux/Demux**	**Paths**	**Disjoint paths**	**Cost = Σ I×O×S**
Banyan [31], Baseline, Reverse Baseline [32], Delta [33], Cube [34], Omega [35], Reverse Omega [32], Shuffle Exchange [36]	8×8	2×2	12/32	**1**	**1**	48
Butterfly [37]	8×8	2×2	32/48	**1**	**1**	96
Dot network [38]	8×8	2×2	13/34	**1**	**1**	52
Gamma [41]	8×8	1×3,3×1, 3×3	32/88	**1**	**1**	192
Monogamma GIN [44]	8×8	1×3,3×1, 3×3	32/88	**2**	**1**	192
Theta [39]	8×8	2×2	10/28/(8/8)	**3**	**1**	72
Scalable crossbar [66]	8×8	2×2	16/64	**1**	**1**	**64**
Augmented SEN (ASEN) [36]	8×8	2×2, 3×3	24/60/**(8/8)**	**4**	**2**	84
2-ary 3-tree (fat-tree) [37]	8×8	2×2,4×4	12/48	**16**	**2**	144
PHI [40]	8×8	2×2, 3×3	10/30/(8/8)	**5**	**2**	82

(Continued)

Table 2.2 Multistage interconnection networks topology comparison. (*Continued*)

Network types	Size	Switch size	Switch/Links/ Mux/Demux	Paths	Disjoint paths	Cost = Σ I×O×S
Extra Stage GIN [42]	8×8	1×3,3×1, 3×3	40/112	**8**	**2**	264
B-network [43]	8×8	3×3,2×2	32/88	**>50**	**2**	208
Fully chained GIN [47]	8×8	2×3,2×1, 3×3	32/112	**64**	**2**	208
Partial chained GIN [47]	8×8	2×1,2×4, 3×2,3×3	32/88	**18**	**2**	200
No Backtracking GIN (NBGIN) [48]	8×8	2×3,2×4, 3×1,3×3	28/80	**4**	**2**	176
Minimal links traversed GIN [50]	8×8	1×4,3×1, 3×3,4×3	32/96	**4**	**2**	224
Combining switches MIN [51]	8×8	2×4,3×2, 3×3,4×4	32/80	**4**	**2**	280
Balanced Gamma [52]	8×8	1×2,2×4, 4×1,4×4	32/96	**4**	**2**	240
Benes [57]	8×8	2×2	20/32	**4**	**2**	80
Extra-Stage Benes [67]	8×8	2×2	24/56	**8**	**2**	96

(*Continued*)

Table 2.2 Multistage interconnection networks topology comparison. (*Continued*)

Network types	Size	Switch size	Switch/Links/ Mux/Demux	Paths	Disjoint paths	Cost = Σ I×O×S
Extra Stage SEN [71] Extra Stage Cube [58]	8×8	2×2	16/40	**2**	**2**	64
Parallel Benes [67]	8×8	2×2	20/80/**(8/8)**	**16**	**2**	112
SEN+2 and Omega +2 [71]	8×8	2×2	20/48	**4**	**2**	80
Kappa network [56]	8×8	1×4,4×1, 4×4	32/112	**8**	**2**	320
Replicated Benes [67]	8×8	2×2	40/96	**8**	**2**	160
Extra group (EGN) [68]	8×8	2×2	12/84 **(12/12)**	**3**	**3**	96
3-disjoint paths GIN [49]	8×8	2×3,2×4, 3×3, 3×1	28/80	**4**	**3**	176
Enhanced IADM [49]	8×8	1×3,3×5, 4×1,5×4	32/112	**7**	**3**	336
Pars network [69]	8×8	2×2	12/108/**(12/12)**	**6**	**3**	120
Improved extra group (IEGN) [68]	8×8	3×3	12/96 **(12/12)**	**12**	**3**	**156**
Extended Generalized SEN [63]	8×8	2×2	64/160/ **(8/8)**	**8**	**4**	320
Incomplete GIN [55]	10×10	1×3, 3×1, 3×3	50/140	**6**	**2**	330

(*Continued*)

Table 2.2 Multistage interconnection networks topology comparison. (*Continued*)

Network types	Size	Switch size	Switch/Links/ Mux/Demux	Paths	Disjoint paths	Cost = Σ I×O×S
Generalized SEN [60]	10×10	2×2	20/50	**2**	**2**	80
Augmented Baseline [59]	16×16	2×2,3×3	16/152/**(16/16)**	**4**	**2**	200
Tandem Omega [62]	16×16	2×2	64/144	**8**	**2**	256
Two-dimensional Ring-Banyan network [61]	16×16	4×4	32/144	**>50**	**3**	512
Clos Network [64]	16×16	4×4	12/32	**4**	**4**	**192**
Modified Baseline [65]	56×56	2×3,2×4, 3×2,4×2	48/192	**1**	**1**	352

Chaining: MINs [47, 61] used chaining links between the stages and switches. But these chaining structure increases the hardware cost and failure probability as any link failure breaks the chain.

Types/Number of switches used: some designs [41–56, 61, 63, 65] are expensive due to the size and number of switches used. Higher hardware complexity of GINs [41–56] is due to 3×3 switches used in place of 2×2 switches.

2.8 Observations

The existing MIN models presented in literature are compared in Table 2.2. These tables provide structure, redundant and disjoint paths, and cost information for all the models. Following observations can be made from the table.

1. Unique-path MINs (Banyan network [31], Baseline [32], Delta networks [33], cube network [34], Omega network [35] and Shuffle-exchange network [36]) provide only one path between any source-destination (S-D) node pairs using 2×2 size switches as basic elements. They lack fault tolerance resulting in poor performance and reliability. Multiple-path MINs provide multiple paths between (S-D) node pairs making them an excellent choice for large-scale multiprocessor systems.
2. Clos [64] is a highly fault tolerant network outperforming other 16×16 networks and four disjoint networks using their modest number of middle-stage switches to minimize implementation cost. They outperform other 16×16 networks.
3. IEGN [68] achieved significant tolerance to faults with relatively low-cost layout compared to other three disjoint networks [49, 69]. The comparison suggests that fault tolerant IEGN topology provides more multiple paths using multiplexers and de-multiplexers at initial and final stages.

4. Replicated Benes [67] and Parallel Benes [67] fault tolerant designs provide two disjoint paths and high reliability in comparison with other fault tolerant networks.
5. Scalable crossbar [66] is a non-blocking network with a very low-cost layout outperforming other single tolerant networks [31–39, 41, 44].
6. Recently developed fault tolerant MINs [60–71] are designed with focus on improving fault tolerance and reliability.
7. Hardware complexity of the design mainly depends on the number and sizes of switches, multiplexers and demultiplexers used. Hardware complexity is found to be higher in multiple-path MINs than unique-path MINs.
8. Gamma Interconnection network (GIN) [41] is a class of MINs providing fault tolerance using redundant paths for transmission of data routed through configurable switches. Various hybrid designs of GIN provide multipath disjoint MINs with higher fault tolerance.

2.9 Summary

MINs are often used as switching fabrics in parallel and distributed systems designed to provide fast and efficient communication between high-capacity processors. This chapter makes an attempt to compare and analyze several recently proposed network topologies of MINs based on some performance metrics. Due to the heterogeneity of metrics, it is difficult to make a direct comparison with other related networks. MINs which provide optimal solution for high connectivity interconnection networks depend only on the usage and types of applications. A survey has been made on MINs for multiprocessor computing and topology comparisons of different MIN architecture are presented.

3

MIN Reliability Evaluation Techniques

3.1 Reliability Performance Criterion

Multistage Interconnection Network (MIN) is an interconnection system consists of multiple layers of interlinked switching elements arranged in a predefined topology to establish communication among processor and memory modules. These networks find applications in wide areas such as telecommunication switching, computer networking, transportation, software structures, integrated circuits, multiprocessor environment, aerospace equipment, multi computers, nuclear power plants, chemical plants and space missions etc.

So, it is necessary to evaluate and improve the performance of these systems through reliability analysis. Reliability is one of the major considerations and important figures-of-merit in design of fault tolerant MINs. Numerous methods had been devised to evaluate MIN reliability [76]. Various reliability measures for evaluating reliability performance of multistage interconnection networks (MINs) are reviewed in this chapter. Relevant terms and definitions related with reliability are presented in forthcoming sections.

Reliability [76, 125] is defined as "The probability that network will successfully perform its intended operations for a specific time under its stated operating conditions". It is the probability that the component or system experiences no failures during the time interval zero to t.

Reliability evaluation [125] is a vital step in all system reliability studies. Based on the literature, metrics used to measure the reliability of MINs are broadly classified into three paradigms as terminal pair reliability, broadcast reliability and network reliability.

Neeraj Kumar Goyal and S. Rajkumar. Interconnection Network Reliability Evaluation: Multistage Layouts, (63–84) © 2020 Scrivener Publishing LLC

3.1.1 Two Terminal or Terminal Pair Reliability (TPR)

The most common communication operation is to send messages from a source node *s* to a terminal node *t*. Terminal pair reliability of a network is defined as the "probability of having at least one operational path/one fault free path between the specified source and destination node". In case of directed networks, it is usually called (*s*, *t*) connectedness.

Now consider the shuffle exchange network with an extra stage (SEN+) for flow of data from source node 0 (000) to destination node 4 (100) as shown in Figure 3.1. Then probability of providing successful connection between these nodes is regarded as terminal reliability evaluation.

3.1.2 Network or All Terminal Reliability (ATR)

Another communication requirement is complete network connectivity that is all nodes are connected with each other. All terminal reliability of a network is defined as the probability that for

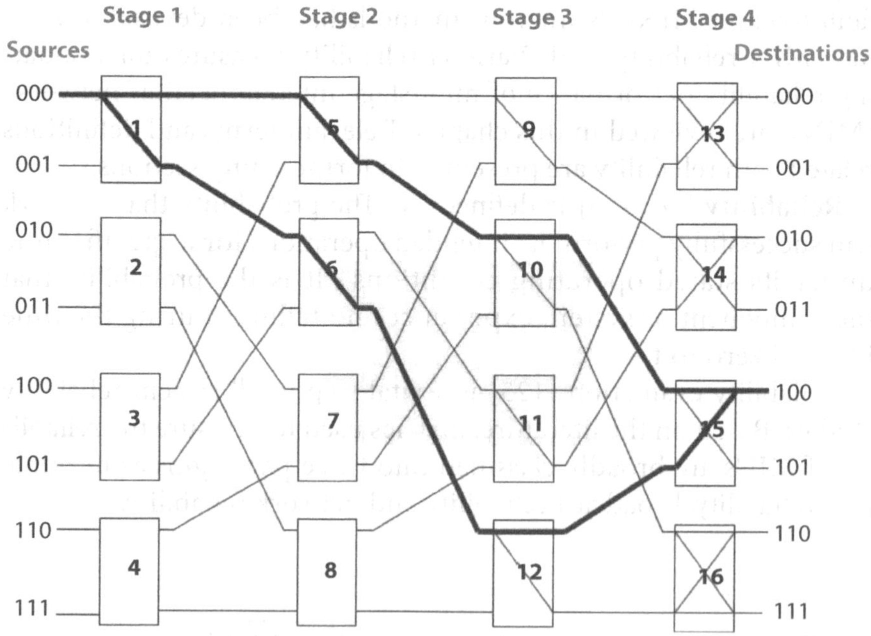

Figure 3.1 Terminal reliability layout of an 8×8 SEN+.

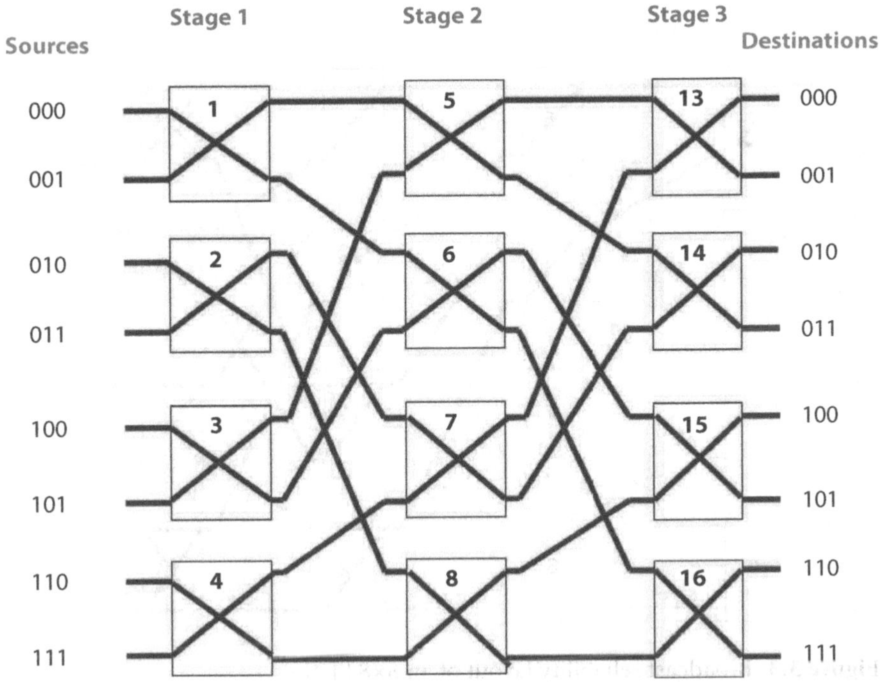

Figure 3.2 Network reliability layout of an 8×8 SEN.

every node pair there exist an operational path to connect them; or equivalently, the probability that there exists a working spanning tree where all nodes of a network can communicate with each other.

Full access capability can be achieved between each input-output pair. Now consider the 8×8 Shuffle exchange network (SEN) network as shown in Figure 3.2, for the flow of data from all source's nodes to all the destinations nodes then the reliability analysis addressing this concern will be termed as network reliability evaluation.

3.1.3 Broadcast Reliability

Another common communication operation is broadcasting. The *broadcast reliability* of a network is defined as the probability of successful broadcast of data communication from a single source node to all the terminal nodes. They are also termed as source-to-multiple terminal reliability (SMT).

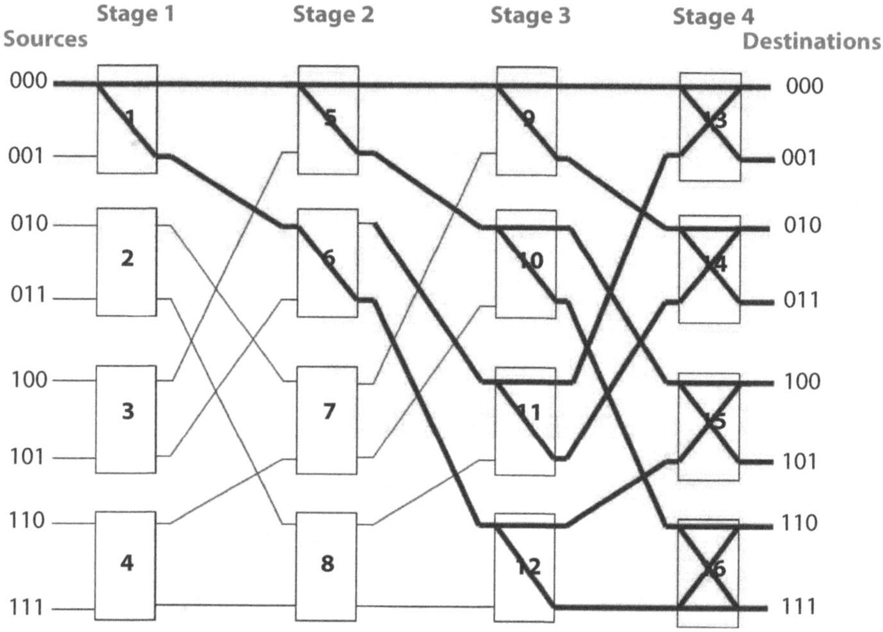

Figure 3.3 Broadcast reliability layout of an 8×8 SEN+.

Now consider the 8×8 shuffle exchange network with an extra stage (SEN+) network as shown in Figure 3.3, for the flow of data from source node 0 (000) to all destination nodes then the reliability analysis between them will be regarded as broadcast reliability evaluation.

3.2 Approaches for Reliability Evaluation

Available literature [77–105] on exact reliability evaluation of MINs is found computationally difficult. Reliability evaluation of MINs has attracted a lot of attention from researchers and many approaches have been presented. A brief study of the few available approaches viz., continuous time Markov chains (CTMC), reliability bounds, reliability block diagram and Monte Carlo simulation are discussed in subsequent sub-sections.

3.2.1 Continuous Time Markov Chains (CTMC)

This method applied for only simple networks where this approach provides exact reliability. Closed-form expressions for the time-dependent reliability (ATR) of the 8×8 and 16×16 shuffle exchange multistage interconnection (SEN and SEN+) networks are computed in [76].

In Markov modeling a set of finite number of states termed as Markov Chain are used to depict the behavior of the system. Different sets of operational switches can be represented by a discrete-state continuous time Markov chain (CTMC) with finite number of states. These states can be divided into two types, a) operational states (Up states) and b) failure states (Down states). We can obtain the network reliability as sum of working states probabilities or 1 – (sum of failure state probabilities). Size of a Markov Chain state space grows exponentially with size of network. For smaller network, CTMC reliability modeling of MIN [76] is very simple but when network size increases, they become arduous and complex due to exponential growth of the state space.

3.2.2 Matrix Enumeration

Matrix enumeration method [79] is listing of all combinations of switching element states (SEs) to calculate the network reliability. Operational status of each combination of switching elements needs to be considered. Network complexity (c) is defined as the total number of MIN switching elements (SEs) and number of possible combinations is 2^c.

3.2.3 Conditional Probability (CP) Method

Conditional probability (CP) method [80] is based on evaluation of every possible switching element states in the network. The actual distribution of working and failed SEs can be analyzed by using conditional probability method. A switching element can only assume either one out of the two states; (i) working or (ii) failed.

Partial functionality loss of switches is considered as failure of entire switch functions. Let working SE state is denoted as '1' (r(t)) and failed SE is denoted as '0' (1- r(t)).

Network reliability based for the above permutation is:

= Stage 1 [1 1 1 1] *Stage 2 [1 0 1 0] *Stage 3 [1 1 0 1] * Stage 4 [1 1 1 1]

$$= [r(t)r(t)r(t)r(t)]* [(r(t). (1- r(t)). r(t). (1- r(t))] * [r(t). r(t). (1- r(t)). r(t))]*[r(t)r(t)r(t)r(t)]$$

$$= r(t)^{4}* (r(t). (1- r(t)). r(t). (1- r(t)). r(t). r(t). (1- r(t)). r(t)) *r(t)^{4}$$

$$= r(t)^{8}. (r(t). (1- r(t)). r(t). (1- r(t)). r(t). r(t). (1- r(t)). r(t))$$

$$= r(t)^{13} (1- r(t))^{3}$$

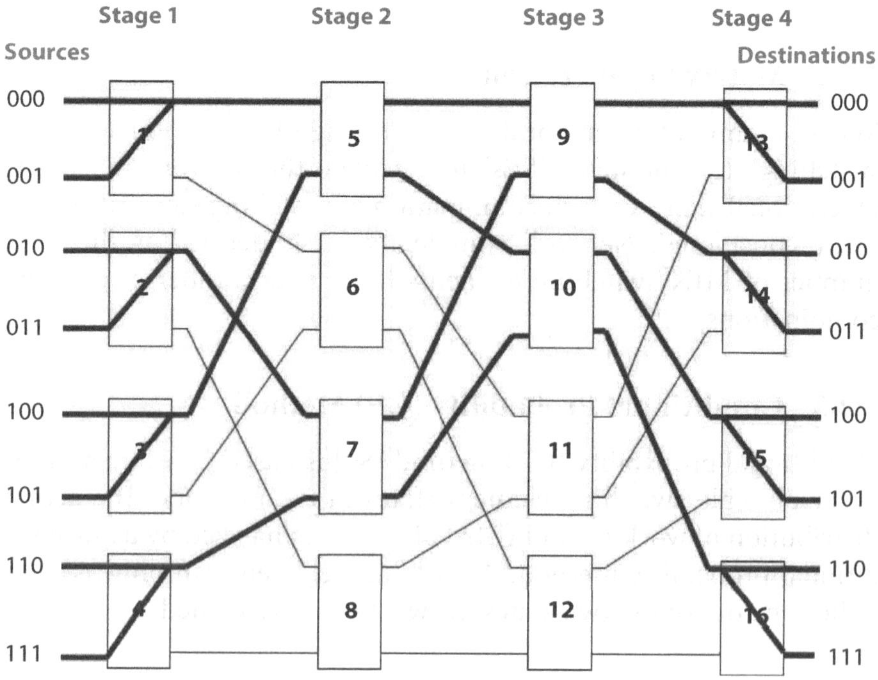

Figure 3.4 Switching element state in an 8×8 SEN+.

Table 3.1 Possible permutation of switching element states for an 8×8 SEN+.

Stage 1				Stage 2				Stage 3				Stage 4			
1	1	1	1	1	0	1	0	1	1	0	1	1	1	1	1

Here working condition of only intermediates stages is considered based on the assumption that switching elements in the first and last stages are perfectly working condition. Consider an example of calculating the network reliability for an 8×8 SEN+. A possible permutation of SE states of Figure 3.4 shown in Table 3.1. Conditional probability (CP) method has been used to derive reliability parameters (terminal, broadcast and network reliability) values specifically for shuffle exchange networks and shuffle exchange networks with extra stages [80].

3.2.4 Graph Models

Graph theory [81] has drawn increased interest of scientists and engineers in the last several decades. Since the evolution of graph theory, it has got a wide range of application in network reliability evaluation. It provides a framework for the graphical representation and mathematical modeling of a complex system or network. According to this concept, a communication system can be modeled as a graph G = {N, L} where N represents the set of nodes or the communication centers and L represents the set of links or connection services.

For example, computer communication network's probabilistic graph nodes [82] represent the computers/switches/transceivers/routers and edges represent the communication link. For reliability analysis, graph presentation of networks is considered to be quite simple, efficient and effective. Therefore, it is used throughout this study. With this graph model, depending on the state (working or failed) of nodes (or vertices) and/or links (or edges), the network can be considered either working or failed.

Multistage interconnection networks are modeled by a probabilistic connected graph [83] where set of switches (N) are connected

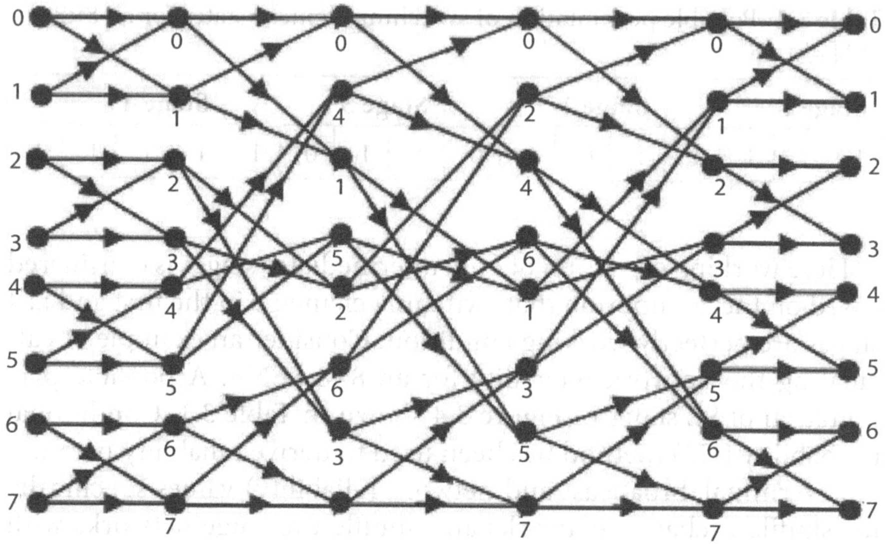

Figure 3.5 Reliability logic graph of 8 x 8 SEN+.

by set of links (L). These graph models are found helpful in comparing the permutation capabilities of various multistage interconnection networks. They provide an easy way of resolving conflicts in designing a suitable MIN. The set of paths as shown in Figure 3.5 for SEN+ network between the given pair of terminals is represented as redundant paths.

3.2.5 Decomposition Method

The method begins by the selection of a key component (K) for its reliability evaluation. The reliability may be expressed in terms of the keystone component based on the theorem of total probability as:

$$R = P(\text{system good} \mid K)\, P(K) + P(\text{system good} \mid)\, P()$$

Where P(system good |K) is the probability that the system is functioning given that K is functioning and P(system good |) is the probability that the system is functioning given that K is not functioning. Terminal reliability of extra stage cube, new four tree multistage interconnection network (NFT), extra stage shuffle exchange network and multi path chained baseline network (MPN) values are computed in [84].

3.2.6 Reliability Block Diagram (RBD)

Reliability block diagram [66–72] can be used to aid the derivation of bound equations which is an approximation technique for determining MIN reliability. System reliability often depends on the reliability of the components used in the system. As a result, we need to have precise information about the components of a system, their reliability values and logical relationships among them from the reliability perspective. A reliability block diagram (RBD) is a visualization that illustrates the simple and redundant arrangement of critical modules that are required to be operational to deliver service.

RBD method presents a tool for modelling a complex MIN structure into a combination of components arranged in different series-parallel pattern. This simple visualization is surprisingly powerful because it enables one to gradually expose the complexity of a system to analysis. Therefore, the RBDs can be very useful for analyzing MINs reliability. Hence RBD is used to model and analyze the reliability relationship of MIN representation from the view point of reliability of components in a system. Each element of a system is represented by a block that is interconnected to other blocks of the system at a desired level of assembly.

The basic relationships among components are depicted as lines that may be series, parallel, series–parallel and complex RBDs as shown in Figure 3.6–3.8. A series system is a configuration that requires all components in working condition for successful operation, failure of single component leads to system failure.

In parallel system, the success of a single component can cause the entire system to be successful. Components of a system can be a combination of series and parallel forms and they are known as the series–parallel RBD.

Consider an example of calculating the terminal reliability (TPR) for an 8×8 SEN+ network given in Figure 3.9. SEN+ terminal reliability can be calculated by visualizing the SEN+ terminal connection topology as two parallel series of SE connections. The expression to calculate terminal reliability using RBD is,

$$R_{Tpr}(t) = [r(t)]^2[1-(1-r(t)^{\log_2 N-1})^2]$$

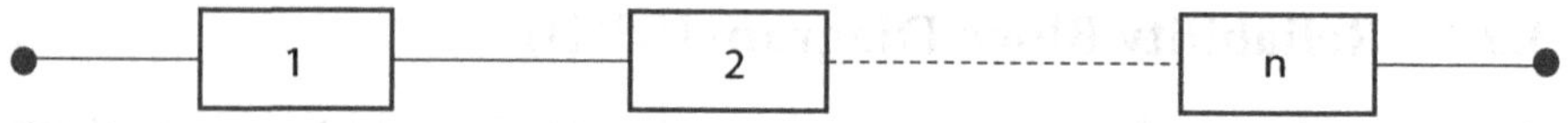

Figure 3.6 Series RBD.

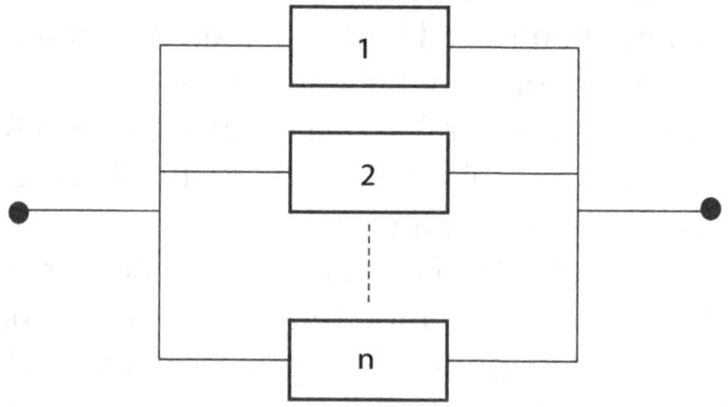

Figure 3.7 Parallel RBD.

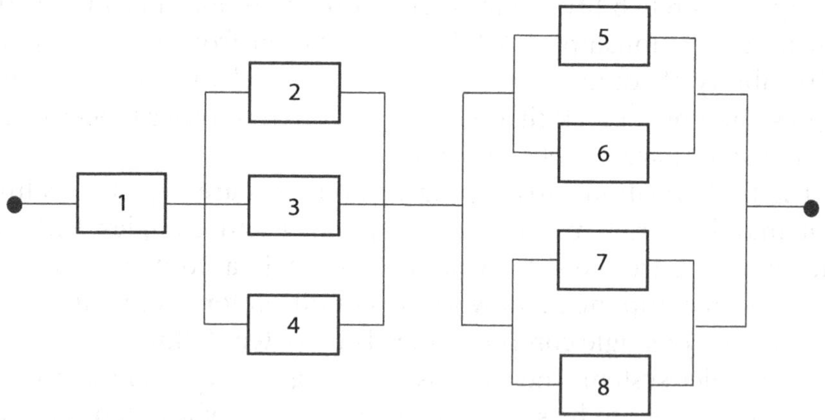

Figure 3.8 Mixed parallel RBD.

Substituting $r(t)= 0.90$, $R_{Tpr}(t) = 0.780759$

The middle stage from Figure 3.9 (Stage 1 to Stage N-1) is a Parallel Configuration and simplify by using conversion form. A parallel structure reliability R_p (t) with non-identical units or components reliability is given by

$$R_{Tpr}(t) = [1-(1-r(t)^{\log_2 N-1})^2]$$

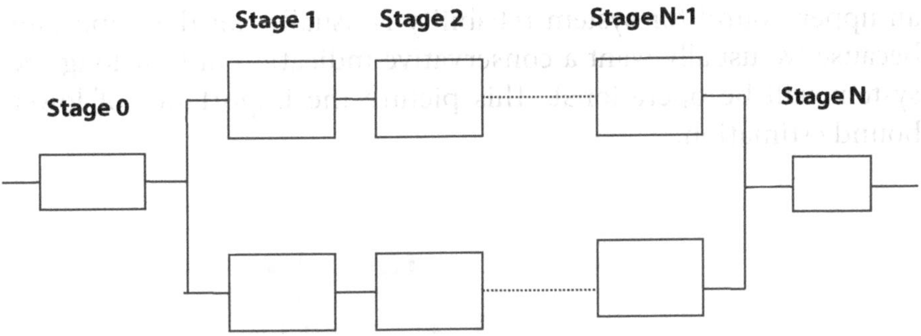

Figure 3.9 Terminal reliability RBD of 8×8 SEN+.

Finally, the expression to calculate terminal reliability using RBD will become

$$R_{Tpr}(t) = \text{Stage 0} * (\text{Stage 1 to Stage N-1}) * \text{Stage N}$$

$$R_{Tpr}(t) = [r(t)] * [1 - (1 - r(t)^{\log_2 N-1})^2] * [r(t)]$$

$$R_{Tpr}(t) = [r(t)]^2 [1 - (1 - r(t)^{\log_2 N-1})^2]$$

Reliability block diagram applied on variety of MINs like Pars network [69], Scalable crossbar network [66], shuffle exchange network [36], Augmented SEN (ASEN) [36], shuffle exchange network with extra stages [71], extra group network (EGN) [68] and improved extra group network (IEGN) [68].

3.2.7 Reliability Bounds

They are also termed as Lower and Upper Bounds (LUB) Method [85–98]. Reliability bound method (lower and upper bounds) is a better alternative to estimate MIN reliability with larger number of inputs. Reliability block diagram can be used to aid the derivation of bound equations of network reliability.

Upper bound value represents the optimistic view of the network to provide the interconnection for a specific period of time. Finding an upper bound for system reliability is usually not the important because we usually want a conservative indication of how long the system will be operational. This picture the importance of lower bound estimation.

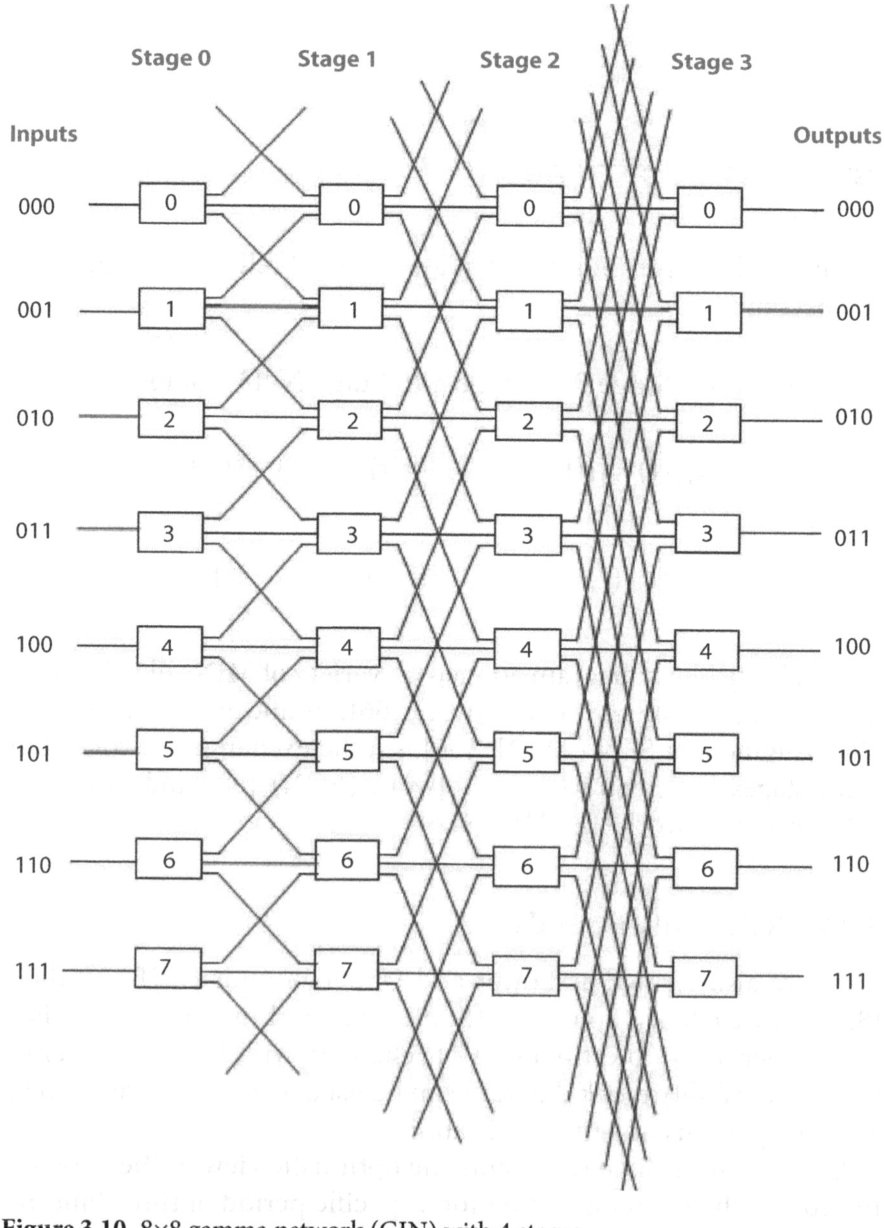

Figure 3.10 8×8 gamma network (GIN) with 4 stages.

Lower bound value [85–90] represents the pessimistic view that the network is able to provide interconnection at least for a specific period of time. Lower bound provides sufficient and reasonable approximation to the actual network reliability such that the system will be operational over the specified time interval.

Consider an example of calculating the network reliability for an 8×8 Gamma network as shown in Figure 3.10.

3.2.7.1 Lower Bound Reliability

Reliability can be can be computed by considering the first and the last stage of the network are series subsystems and the two middle subsystems are parallel-series subsystems. The lower bound network reliability is shown in Figure 3.11 and expressed as:

- $R_{Nr\text{-}Lb}(t)$= Stage 0 * (Stage 1 to n-2) * Stage (Stage 1 to n-1) * Stage n
- $R_{Nr-Lb}(t)=[r(t)]^{N}*[1-[1-r(t)^{N/3[(\log_2 N)-2]]3}]*[1-[1-r(t)^{N/2]}]*[r(t)]^{N}$
- $R_{Nr-Lb}(t)=[r(t)]^{2N}*[1-[1-r(t)^{N/3[(\log_2 N)-2]]3}][1-[1-r(t)^{N/2]2}]$
- Substituting r(t) = 0.99, $R_{Nt-Lb}(t) = 0.785664$

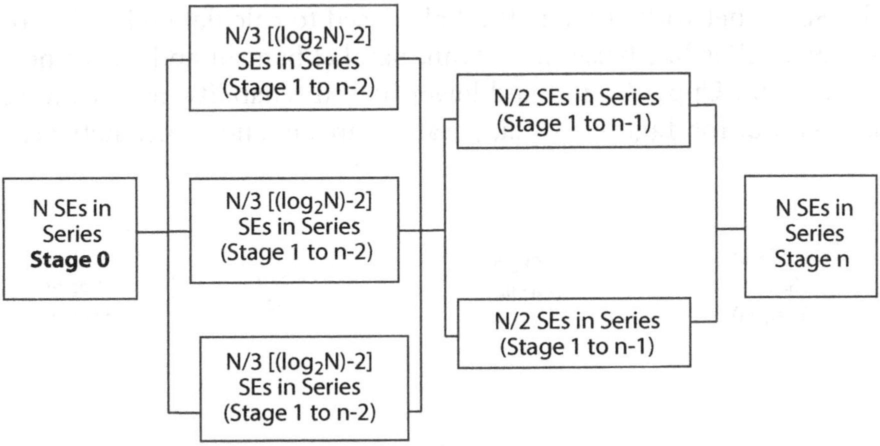

Figure 3.11 Lower bound reliability of N×N gamma network.

3.2.7.2 Upper Bound Reliability

Upper bound reliability [90–98] can be obtained by assuming that the first and last stage of network of N switches are represented in series and the two intermediate stages of N switches arranged in parallel as shown in Figure 3.12. The network is still functioning provided that only one each intermediate stage switch must be operational for successful communication for all source–destination pairs. The upper bound network reliability is expressed as:

- $R_{Nr\text{-}Ub}(t)$= Stage 0 * Stage (Log_2 N) –1 * Stage n
- $R_{Nr-Ub}(t)=[r(t)]^{2N}[1-[1-r(t)]\,N]^{(\log_2 N)-1}$
- Substituting r (t)= 0.99, $R_{Nt\text{-}Ub}(t)$ = 0.851458

Initial bounds calculation for MINs was introduced by Blake and Trivedi [78]. They derived slightly tighter lower and upper bounds on network reliability for large networks in SEN and SEN+.

Then bounds method was applied to calculate reliability of interconnected parallel banyan network (IPBN) and the circular IPBN (CIPBN) [88] with the help of reliability block diagram. Mean time-to-failure (MTTF) for both IPBN and CIPBN also derived [89]. MTTF is defined as the expected time elapsed before some sources are disconnected from some destination.

Further improvements are made [23] in deriving broadcast reliability and more efficient tight bounds for the network reliability of the SEN+ network. This method also used to calculate reliability of various MINs like baseline banyan, parallel banyan and Benes networks [90]. Upper bound and lower bound reliability analysis were carried out for Tagle-Sharma, parallel banyan, and Benes networks

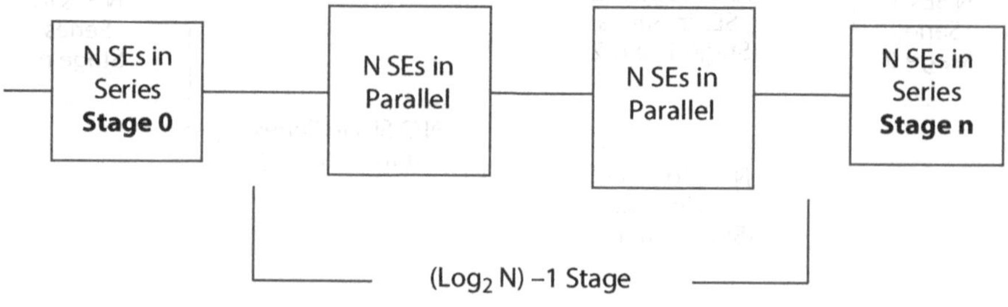

Figure 3.12 Upper bound reliability of N×N gamma network.

[88, 89]. Bounds methods was also incorporated to calculate reliability of networks like INDRA, extra stage cube, omega networks and Mux–Demux replicated MIN [91].

Reliability analysis of various irregular MINs has been presented and compared in [92]. The networks include FT (Four Tree), MFT (Modified Four Tree), NFT (New Four Tree), IFT (improved Four Tree), IASN (Irregular Augmented Shuffle) and IIASN (Improved Irregular Augmented Shuffle). Reliability studies by Sharma [98] show that their proposed irregular modified augmented baseline network (IMABN) achieved a better performance over modified augmented baseline network (MABN). The bounds on reliability are computed by Nitin [97] for Zeta network, Phi Network and Quad-tree Networks (QTNs).

Gunawan derived lower bound and upper bound evaluation of large network reliability for various MIN networks. He derived reliability bounds in shuffle-exchange [23], extra-stage cube network and gamma networks [85–87]. Gunawan showed that lower bound reliability of the network provides a close estimate of the actual network reliability for extra-stage cube network.

This method is widely used in recent literature. Some of the surveyed existing networks [93–97] includes: Modified four tree network (MFT-2), fault-tolerant advanced omega network (FTAON), modified augmented shuffle exchange network (MASN), Augmented Shuffle Exchange Network (ASEN-2), irregular augmented shuffle exchange network (IASEN), improved four tree network (IFTN), irregular augmented baseline (IABN) and augmented baseline (ABN).

3.2.8 Monte Carlo Simulation

Monte Carlo simulation approach [99] is statistically sound and easy to understand, however it is computationally expensive and experimental problem-solving technique.

Monte Carlo analysis is a powerful tool for modeling the reliability of complex systems. Proper application of this tool requires understanding of its underlying principles. When applied to reliability analysis, Monte Carlo simulations [100] use the failure and repair statistical distributions of individual units to model the system behavior over time. The simulation results can then be used to make more accurate decisions for improving system reliability.

To perform the simulation, a random number (RN) between 0 and 1 is generated to represent the probability of an event occurring at a given time. From the probability, a corresponding event-time can be calculated using the statistical distribution chosen to model the process. Reliability estimation [101] gives statistical estimates by performing random sampling of switching element (SE) states. Monte Carlo simulations require large sample size to obtain a sufficient quantity of failures to provide reliable estimate. The number of replications or trials represents the total number of random sampling needed in the MC method, which is also known as the sample size.

Because reliability of each SE is based on probability distributions, they can be modeled by a set of random numbers. For instance, if the reliability of a SE is 0.9, then working SE can be represented by the numbers below 0.9 and failed SE by the numbers from 0.9 to 1.0.

Therefore, each SE [102] state can be simulated by generating random numbers. For each simulation, it is checked whether particular set of conditions lead the network to working or failed state. As the number of simulations increases, the sample mean of these independent estimates approaches the actual characteristics of the system.

In case of MINs, analyzing the switching element (SE) states by randomly generate samples for the switching stages and then evaluate its connectivity. Accuracy of this estimation can be increased by simply increasing the number of samples. However, increasing number of replications will increase required computational resource and time substantially required for generating the samples.

Estimation and enumerating all the possible combinations using Monte Carlo method can become time consuming and rather complex when the network size increases. Therefore, MIN estimation methods to evaluate reliability of larger networks rely on other methods. Monte Carlo method [103, 104] has been used to derive reliability parameters (terminal, broadcast and network reliability) values specifically for shuffle exchange network with additional stage (SEN+) topology and extra-stage cube (ESC) network.

3.2.9 Path-Based or Cut-Based Approaches

The path-or-cut based approaches require path set or cut set information of the network [125] as the starting point of the reliability evaluation process. A vast majority of methods to evaluate network reliability belong to POC paradigm. These techniques work in a two-step procedure viz.

1. Enumeration of minimal path sets or cut sets of the network, and
2. Combining path sets or cut sets logically, using the laws of union and idempotence.

In literature, a number of approaches are presented to carry out these two steps. Both of these steps are not as simple as they may appear to be, particularly in case of large and interconnected networks, which could be *planar* or *non-planar*. First of all, we must have efficient algorithms to enumerate path sets or cut sets, which itself may sometimes become an enormous task.

Further, in a highly interconnected network, the number of path/cut sets is bound to be quite high. The other stumbling block in the procedure of TPR evaluation is unionizing of path sets or cut sets to obtain a TPR expression.

This can also be quite cumbersome and unwieldy task. This unfavorable situation is due to computational difficulties in unionizing the events inherent to the problem, which increases exponentially with network complexity.

Therefore, to make network reliability computation [106–120] using path or cut sets competitively attractive a lot of research has been carried out in order to develop:

1. A simple and fast method of enumerating all path or cut sets and
2. An efficient method to handle these path sets (or cut sets) to compute network reliability expression.

Reliability evaluation of multistage interconnection networks is generally achieved by enumerating path sets or cut sets of the network. A path set [125] is defined to be a set of minimal paths connecting source and destination node. A path is set of components whose functioning ensures that the system functions. A cut set is defined to be a set of minimal cuts that disconnects the source and destination nodes. A cut is a set of components whose failure will result in system failure.

Various techniques have been employed for enumerating the path or cut sets of a network graph and computing network reliability expression from this path or cut sets. Several methods (Aggarwal *et al.* [86], Aziz *et al.* [106], Beigel [109], Danielson [110], Fratta and Montanari [111], Krishnamurthy and Komissar [112], Misra and Rao [113, 114], Rai and Aggarwal [115], Samad [116]) are available in literature for the enumeration of path sets. These methods are based on symbolic solution, linear equations, power of connection matrix, expansion or reduction of adjacency matrix and rearrangement of path sets. A multi-way tree structure-based method for enumeration of path set was proposed by Fard and Hunhee [117].

Abraham method (Single Variable Inversion) also called as Recursive disjoint product was introduced by Abraham [86]. An approach wherein variables are inverted sequentially one at a time. Hiedtmann method (Multiple Variable Inversion) KDH 88 is an improvement over SVI technique. In contrast to SVI approach wherein variables are inverted one at a time in a sequence, in MVI approach [121] a group of variables are inverted at once. Using the min path sets, it recursively generates the terms for each path set, which are disjoint to all prior path sets. For example, if we have two paths $\{x_1, x_3\}$ and $\{x_2\}$ disjoint expression using MVI be:

$$\text{First term: } x_1 x_3;$$

$$\text{Second term: } \overline{x_1 x_3} x_2$$

And the structure function is

$$\Phi(X) = x_1x_3 + = \overline{x_1x_3}x_2 = x_1x_3 + (1 - x_1x_3)x_2$$

Accurate reliability computations of multistage interconnection networks are recently computed by using these methods. Recently path set enumeration-based sum of disjoint products (SDP) approaches used in computing the reliability of parallel computing systems. The reliability and total number of redundant paths for MINs are

Table 3.2 Terminal reliability expression of 8×8 gamma networks.

Tag value (δ)	Number of paths	Terminal reliability
0	1	r^4
1	4	$3r^4 - 3r^5 + r^6$
2	3	$2r^4 - r^5$
3	5	$3r^4 - 3r^5 + r^6$
4	2	r^4
5	5	$3r^4 - 3r^5 + r^6$
6	3	$2r^4 - r^5$
7	4	$3r^4 - 3r^5 + r^6$

Substituting r (t) = 0.99, Rδ = 4(t) = r4 = 0.994 = 0.9605.

calculated by this efficient methodology which first enumerate minimal path sets and then evaluate reliability expression in compact form of sum of disjoint products (SDP) using improved multi-variable inversion (MVI) algorithm.

Consider an example of calculating the terminal reliability for an 8×8 Gamma network shown in Figure 2.16. Enumerating path sets for the network which gives different paths for different tag values and different reliability equations and shown in Table 3.2.

3.3 Observations

The following observations can be made from the literature survey covered in the previous section and Issues with existing survey and motivation for research work are:

1. Various algorithms and methods, stated in literature consider only connectivity as the network success criteria.
2. All the previous literature considers links to be perfect. In these systems perfect nodes are not realistic as the nodes tend to have high probability of failure due to high complexity.

3. Most of the MIN reliability evaluation existing in literature used reliability block diagram and bound equations methods provides approximate reliability values. They are usually approximation technique used to determine the overall system reliability. But when complexity of MINs increases the evaluated reliability values varies.
4. A vast majority of reliability evaluation methods like Monte Carlo simulation, graph models, conditional probability method, matrix enumeration and continuous time Markov chains focused only on shuffle exchange networks (SEN), Omega networks and cube networks which are single fault tolerant networks.
5. Most approaches in the literature provide only better approximation values, which vary for the same network. However, there is always a demand in finding the exact reliability of multistage interconnection networks for both small and large scale complex parallel interconnection systems.
6. Most literature calculated only terminal pair reliability values. Very less research work has been done on broadcast reliability evaluation.
7. All terminal reliability values are calculated based on bounds method only. Lower bound provides approximation to the network reliability such that the system will be operational over the specified time interval. But the exact reliability needs to be evaluated to avoid degradation of the network performance.
8. Very less research work has been done on the reliability evaluation of networks belongs to GIN family, Kappa network and two-dimensional ring-banyan network due to their complicated structural construction. Their exact reliability values need to be evaluated.
9. Individual literature calculated reliability parameters (TPR/ATR/Broadcast) for the less fault tolerant networks only. However, literature lacks evaluating all the reliability parameters for major and complex MINs.
10. Reliability evaluation of major recently developed MINs is yet to be carried out.

11. There is no mention of any definition and/or enumeration of disjoint path information of MINs. Disjoint paths give better information about network capability to withstand minimum number of switch failures.
12. The non-path-based approaches tend to be less efficient as compared to path based algorithms for complex networks. Path set based approach provide accurate reliability computation of MINs.
13. Approaches reviewed above considered single node pair requirements only, but networks need to evaluate the reliability of fulfilling demands appearing across multiple node-pairs simultaneously. Surprisingly there is no literature that evaluated MIN reliability based on multiple sources and multiple destination nodes.
14. Comprehensive MIN reliability analysis is needed to evaluate the reliability of MIN to allow comparison be made against other MIN topologies in order to select the best topology that could meet the reliability requirement.

In this chapter a global view of the different types of reliability measures and different approaches for the evaluation of these reliability measures is presented. A comprehensive review of the various approaches present for evaluation of different reliability measures is also presented. Based on the literature, different shortcomings have been discussed and work objectives have been identified.

4

Terminal Reliability Analysis of MIN Layouts

Multistage Interconnection Networks (MINs) are widely used for reliable data communication in tightly coupled large-scale multiprocessor systems. Reliability is a key parameter [85–90] in the design and operation of modern large and complex multistage interconnection networks (MINs). Accurate reliability evaluation of interconnection networks is still a challenge due to modeling and computation complexities.

Components of these modern complex MINs are apt to random failures. Hence, for safe and economically fruitful operation, reliability evaluation of these complex networks has become more and more important. Need of reliability measures for MINs is quite evident as these measures provide performance assessment from user point of view.

The effective and creative MIN design approach needs some careful thinking on the part of design engineer to create a new or improved system with high reliability. Terminal pair reliability (TPR) is the most commonly used reliability performance index of multistage interconnection networks. The network reliability is usually concerned with the task of evaluating the *terminal reliability* or the probability of establishing communication between two specified nodes usually called as *source* and *sink*.

TPR evaluation [92–94] considers only availability of connection (path) from single source node to single destination node and is defined as the probability that there is at least one operational path between the source and destination node. To evaluate two-terminal reliability (2TR) of future envision reliable MIN design, reliability analysis of current existing MIN Layouts is essential. Available

Neeraj Kumar Goyal and S. Rajkumar. Interconnection Network Reliability Evaluation: Multistage Layouts, (85–114) © 2020 Scrivener Publishing LLC

literature on reliability evaluation [19, 79, 80] of MINs suggest use of various methods like decomposition method, reliability block diagram, Continuous time Markov chains (CTMC), reliability bounds, Monte Carlo simulation etc. However, these methods failed to provide exact reliability of multistage interconnection networks for both small and large scale complex parallel interconnection systems due to computational complexity or being approximate in nature.

Aiming to evaluate exact reliability, this chapter presents simple and computationally efficient approach given by Chaturvedi and Misra [124]. This approach is based on enumerating minimal path sets of a network and calculating reliability from these minimal path sets. Network reliability computation using path set approaches are computationally efficient. Therefore, reliability evaluation is carried out using path sets of a probabilistic graph in this work. Methodology can be divided in following two major steps:

- A simple and fast method of enumerating all path sets and
- An efficient method (Hybrid multi-variable inversion (MVI) algorithm) to compute network reliability expression from these minimal path sets.

Major contributions of this chapter are summarized as follows:

- Efficient approach for enumerating path sets and evaluating exact reliability of MIN. The method is supported by well explained examples at each step.
- To prove the efficacy and correctness of the applied approach, approach is compared with other reliability evaluation approaches by using a common MIN (shuffle-exchange network (SEN)). Terminal, broadcast, and network reliability for the shuffle-exchange network (SEN), SEN with an additional stage (SEN+), and SEN with two additional stages (SEN+2)) systems are analyzed and compared with the other approaches.
- Then we extend our work to evaluate and compare various benchmark MINs by calculating their terminal pair reliabilities.

4.1 Chaturvedi and Misra Approach

We need an efficient method for enumerating path as these path sets are used to obtain system reliability expression in a compact form.

Several methods (Aggarwal *et al.*, [86], Aziz *et al.*, [106], Beigel, [109], Danielson [110], Fratta and Montanari [111], Krishnamurthy and Komissar [112], Misra and Rao [113, 114], Rai and Aggarwal [115], Samad [116]) are available in literature for the enumeration of path sets. These methods are based on symbolic solution, linear equations, power of connection matrix, expansion or reduction of adjacency matrix and rearrangement of path sets. A multi-way tree structure-based method for enumeration of path set was proposed by Fard and Hunhee [117].

Chaturvedi and Misra [124] proposed a data-structure-array based minimal path set enumeration approach. The techniques based on *sum of disjoint products* (SDP) are often used effectively and efficiently in TR evaluation to provide a compact TR expression from these minimal path sets. These techniques start with a Boolean polynomial formed by either success terms (path sets) or failure terms (cut sets). This class of techniques relies on Boolean algebra manipulation to convert these terms into a set of *exclusive and mutually disjoint* terms. The algebraic sum of these terms provides reliability expression.

Both the SVI and MVI techniques accept a Boolean function of network variables (nodes and/or links) representing minimal path sets or cut sets to compute network reliability. However, in contrast to SVI approach wherein variables are inverted sequentially one at a time, in MVI approach a group of the variables are inverted at once. In SVI, the minimization of the expression has been achieved through different ordering schemes of path sets or cut sets so far.

In MVI, not only the ordering of path sets or cut sets but the simultaneous inversions of a group of variable results yield a more compact expression as compared to the expression that would be rendered by an SVI technique. This provides fewer numbers of disjoint terms, each of which covers a greater domain of Boolean structure function. In other words, the MVI techniques have a built-in feature of rendering a compact reliability expression.

However, with invent of MVI technique, the SVI now have a lesser weight. The MVI-SDP based method was first proposed by Heidtmann (1989). Later, Veeraraghavan and Trivedi (1991) proposed another algorithm for the symbolic reliability analysis of a distributed system with s-independent components.

Further improvement and simplification on this approach 'CAREL' is presented by Soh and Rai [118] extending the method of Rai and Aggarwal [120]. By taking the advantages of Soh and Rai [118, 119] method, Chaturvedi and Misra [124] presented hybrid methods and have shown that these methods perform even better than (Heidtmann [121] and Soh [118]) in terms of the lesser number of SDP terms generated and CPU execution time.

Therefore, to make network reliability computation using path sets competitively attractive and this approach has been carried out by:

1. A simple and fast method of enumerating all path sets and
2. An efficient method to handle this path sets to compute network reliability expression.

4.1.1 Path Set Enumeration

This section emphasizes on the path-based approach for reliability evaluation of MINs. Definitions of some important terms used in this work are presented in this section.

Multistage interconnection network is modeled by a probabilistic directed graph. A probabilistic directed graph, consists of nodes representing communication centers and edges representing links between the communication centers, is used for representing MINs.

This directed graph, or digraph, is a connected graph, in which every branch has an orientation, between the specified sources and sink nodes. Source node has no branch incident to it and the sink node has no branch incident from the node. Directed edges are referred to as arcs. Arcs represent one-way communication links between two nodes with communication taking place in the direction pointed by arc.

Multistage interconnection network digraph $G = (V, L)$ consists of a set of nodes or vertices V and a set of arcs L. In reliability engineering, a path set is a set of links such that if every link in this set were good, it would result in system success. A minimal path set is a sub set of a path set with minimum number of links that still make the system functional.

Multistage interconnection networks (MINs) can be represented by connection or adjacency matrix. The most popular and versatile form in which a network is represented or fed to a computer is its adjacency matrix. This matrix is being used to test the connectivity criterion of a specified node set.

The assumptions adopted in present study are:

- A Switching elements (SEs) can only have two states; working = 1 or failed = 0
- All SE failures are random and statistically independent.
- SEs is assumed to be less reliable than the link and cannot be repaired.
- All SEs have identical reliability (for calculation purpose only). The proposed approach can take non-identical reliability values also.

This simple approach has favored us to devise methods to look for all path sets between any specified pair of nodes of a given network viz., source and sink. The method to enumerate path sets starts with the generation of adjacency matrix from the network data.

Once the adjacency matrix of the graph is available, the search starts with the source node and builds up path sets successively by selecting row(s) of adjacency matrix corresponding to adjacent node. It looks for nonzero elements in the adjacency matrix in the row corresponding to this node, appends each nonzero entry separately in a new vector, where number of new vectors would be equal to the number of nonzero elements found in the search. It follows the same procedure for the last element added in each of these vectors, by selecting the row of adjacency matrix corresponding to this most recent entry till the sink node is met and thus builds several pathsets in an ordered manner.

Various steps involved in the path set enumeration can be stated as follows:

1. First prepare adjacency matrix of the network.
2. It begins with the source node by looking for nonzero elements in the adjacency matrix in the row corresponding to the source node.

3. It appends each nonzero entry found at a certain position in that row with source node number separately in a new vector.
4. The number of new vectors thus formed would be equal to the number of nonzero elements found in this search.
5. The same procedure is followed for the row corresponding to the last element added in each of these vectors and we select the row of adjacency matrix corresponding to the most recent added entry.
6. This continues until the sink node is met and then path sets are traced back.

Let us take a simple example ARPA network (Fard and LEE [117]) to illustrate the concept:

Illustration 4.1
The ARPA network has been shown in Figure 4.1 for undirected network. The source has assigned the number 1 while the destination is given the number 5. The adjacency matrix is shown in Table 4.1.

Let us consider Figure 4.1 to enumerate path sets. The steps to follow are shown in a tabular form in Table 4.1 and Table 4.2 for the sake of brevity. We start with source node, 1, and store it in a matrix [P]. Matrix [A] is the row of adjacency matrix corresponding to the source node. [PP] contains [P] with each entry of [P] appended by the non-zero's locations of [A].

As A has non-zeros at locations 2 and 3, we append [P] with these values and stored it in [PP]. We take out rows of adjacency matrix

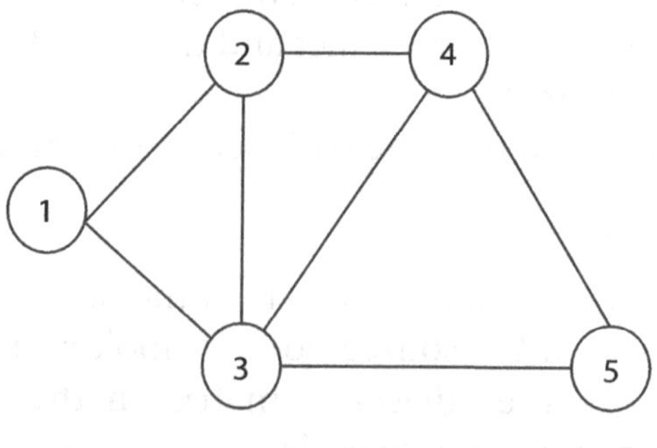

Figure 4.1 Undirected ARPA network.

Table 4.1 Adjacency matrix.

Nodes	1	2	3	4	5
1	0	1	1	0	0
2	1	0	1	1	0
3	1	1	0	1	1
4	0	1	1	0	1
5	0	0	1	1	0

Table 4.2 Steps to enumerate path sets for the network of Figure 4.1.

Iteration no.	[P]	[A]	[PP]	[AA]
1.	1	0 1 1 0 0	1 2 1 3	0* 0 1 1 0 0* 1 0 1 1
2.	1 2 1 3	0 0 1 1 0 0 1 0 1 1	1 2 3 1 2 4 1 3 2 1 3 4 ***1 3 5***	0* 0* 0 1 1 0* 0* 1 0 1 0* 0 0* 1 0 0* 1 0* 0 1 ***0* 1 0* 1 0***
3.	1 2 3 1 2 4 1 3 2 1 3 4	0 0 0 1 1 0 0 1 0 1 0 0 0 1 0 0 1 0 0 1	1 2 3 4 ***1 2 3 5*** 1 2 4 3 ***1 2 4 5*** 1 3 2 4 1 3 4 2 ***1 3 4 5***	0 0* 0* 0 1 ***0 0 0* 1 0*** 0* 0* 0 0* 1 ***0 0 1 0* 0*** 0 0* 0* 0 1 0* 0 0* 0* 0 ***0 0 0* 0* 0***
4.	1 2 3 4 1 2 4 3 1 3 2 4	0 0 0 0 1 0 0 0 0 1 0 0 0 0 1	**1 2 3 4 5** **1 2 4 3 5** **1 3 2 4 5**	

corresponding to the last element in each row of [PP] i.e., 2nd and 3rd rows by setting elements (1 & 2) of 2nd row and elements (1 & 3) of 3rd row to zero (marked with star). We store these modifications in rows in matrix [AA]. It completes one cycle of the search.

As a next step, we update [P] by [PP] and [A] by [AA]. At this point, we perform two checks viz., (i) whether the last element in any row(s) of [PP] contains sink node and, (ii) whether all entries in that row(s) of [AA] are zero. Depending on the observations, we draw the following conclusions:

If last value in some row(s) of [PP] is the sink node number, then path set(s) has been found and we remove this row(s) from [PP] and [AA] before updating them into [P] and [A]. In this case, row(s) of [AA] might or might not have all values zero e.g., it could be seen in iteration number two for path set {1, 3, 5}, and in iteration number three for path set {1, 3, 4, 5} and are shown in bold and italic.

However, if some row(s) of [AA] has all zeros but [PP] does not have last element as sink node number, this implies formation of loop(s). This can be seen for the set {1, 3, 4, 2} in iteration number three, which is shown in bold and underlined. Hence, we discard this set(s) also and proceed further with the remaining entries in [PP] and [AA]. We stop the procedure when the iterative count becomes one less than the number of nodes in the network. Thus the minimal path sets of the above example would be:

{1, 3, 5}, {1, 2, 3, 5}, {1, 2, 4, 5}, {1, 3, 4, 5}, {1, 2, 3, 4, 5}, {1, 2, 4, 3, 5} {1, 3, 2, 4, 5}

The following points may be noted for the illustration outlined above:

1. Size of [AA] is fixed and equal to the nodes in the graph.
2. Maximum size of [PP] = total non-zeroes entries in [A]* Iteration number (i.e., cardinality of the path)
3. At any point of time, the length of P would be equal to the (iteration count +1).
4. If at any point, if a particular row of [PP] contains sink node number then it is a valid path set.
5. In [AA] if all elements in a row are zero and last node in PP is not sink node number then it is an Invalid path set. Otherwise a valid path set.

The highlights of the method presented above may be summarized as follows:

- It can find path sets for a specified pair of nodes.
- The iterative count also depicts the cardinality of a path set, if found.
- It can also convert path set in terms of network nodes to network links.

Illustration 4.2

We consider a delta network (general MIN) [33] as shown in Figure 4.2. This 8×8 network is generally represented having 8 sources and 8 destinations connected by **12 switches** through **3 stages.**

The network is represented or fed to a computer is its connection matrix. Adjacency matrix is 'N×N' matrix, whose N rows corresponds to N nodes (switches in MIN) of the network. Table 4.3 shows the adjacency matrix for the network in Figure 4.2.

Any row having '1' in any column of the Adjacency matrix indicates that two nodes are interconnected by a link. For example, 6th row of Table has value '1' in the columns named 9, 10 which interprets

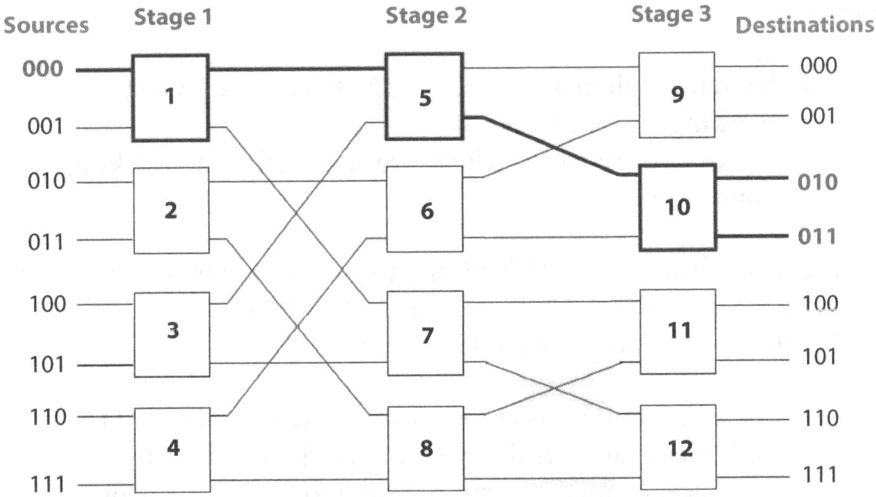

Figure 4.2 8×8 delta network.

Table 4.3 Adjacency matrix for the network in Figure 4.2.

	1	**2**	**3**	**4**	**5**	**6**	**7**	**8**	**9**	**10**	**11**	**12**
1	0	0	0	0	1	0	1	0	0	0	0	0
2	0	0	0	0	0	1	0	1	0	0	0	0
3	0	0	0	0	1	0	1	0	0	0	0	0
4	0	0	0	0	0	1	0	1	0	0	0	0
5	0	0	0	0	0	0	0	0	1	1	0	0
6	0	0	0	0	0	0	0	0	1	1	0	0
7	0	0	0	0	0	0	0	0	0	0	1	1
8	0	0	0	0	0	0	0	0	0	0	1	1
9	0	0	0	0	0	0	0	0	0	0	0	0
10	0	0	0	0	0	0	0	0	0	0	0	0
11	0	0	0	0	0	0	0	0	0	0	0	0
12	0	0	0	0	0	0	0	0	0	0	0	0

that the node 6 is connected with node 9 and node 10. Thus, in this network,

- Switching elements (1, 2 ....12) have only two states; working = 1 or failed = 0
- Links are reliable. They are perfectly in working condition

The algorithm starts with building the adjacency matrix as shown in Table 4.3 and path sets are enumerated in Table 4.4.

The iterations start from step 2 onwards.

1. As A has non-zeros at locations 5 and 7, we append [P] with these values and stored it in [PP]. We take out rows of adjacency matrix corresponding to the last element in each row of [PP] i.e., 5th and 7th rows by setting elements (1 & 5) of 5th row and elements (1 & 7) of 7th row to zero

Table 4.4 Steps to enumerate path sets for the network of Figure 4.2.

Iteration no.	[P]	[A]	[PP]	[AA]
1.	1	000010100000	1 5	000000001100
			1 7	000000000011
2.	1 5	000000001100	1 5 9	000000000* 100
			1 5 10	***0000000010* 00***
	1 7	000000000011	1 7 11	00000000000* 1
			1 7 12	000000000010*

(marked with star). We store these modifications in rows in matrix [AA]. It completes one cycle of the search.

2. As a next step, we update [P] by [PP] and [A] by [AA]. At this point, we perform two checks viz., (i) whether the last element in any row(s) of [PP] contains sink node and, (ii) whether all entries in that row(s) of [AA] are zero.
3. If last value in some row(s) of [PP] is the sink node number, then path set(s) has been found and we remove this row(s) from [PP] and [AA] before updating them into [P] and [A].
4. Thus the minimal path sets of the above example would be: {1, 5, 10}

The similar steps could be followed for other families for MINs to enumerate path sets. The algorithm can be applied to general network-graph containing undirected, directed or both types of branches.

In a nutshell, the methods presented in this chapter are found to be a simple and efficient method and is computationally economic in comparison to several other methods described in the literature for enumerating minimal path of a network. Thus, this path set enumeration approach applied to variety of MINs and reliability values are evaluated from these path sets by using hybrid MVI algorithm.

Next section discusses the reliability evaluation approach using path sets generated in this section.

4.1.2 Reliability Evaluation using MVI Techniques

Generally, for large and complex Multistage interconnection networks (MINs) systems, *terminal reliability* (TR) evaluation [106–110] is quite a cumbersome and time-consuming process due to the large number of terms in the resulting reliability expression.

Thereafter, the process of assessing numerical value from the obtained expression makes evaluation of accurate value of system reliability a difficult job due to buildup of round-off error. This restricts the use of a low-end PC for computing reliability of such networks.

The methods based on *sum of disjoint product* (SDP) [118–124] technique have been used to evaluate a compact and efficient *terminal reliability* expression. In fact, with the advent of multiple inversions techniques, a new impetus has been given to the techniques of network reliability evaluation using path/cut sets of the network to produce a minimized form of expression. Compact reliability expression helps in tackling larger multistage interconnection networks (MINs).

In single variable inversion approach, variables are inverted sequentially one at a time. However, multiple variable inversion approach is an improvement over SVI technique. In MVI approach a group of variables are inverted at once. Using the min path sets, it recursively generates the terms for each path set, which are disjoint to all prior path sets.

Let us take a simple example to illustrate the concept to get disjoint expression using MVI concept:

Illustration 4.3

The sample network considered is shown in Figures 4.3 contains 5 nodes and 8 links.

The enumeration methods described in Section 4.1.1.2 enumerates 9 path sets for this network of Figure 4.3 as follows:

{(1,6), (2, 7), (4, 8), (1, 3, 7), (2, 3, 6), (2, 5, 8), (4, 5, 7), (1, 3, 5, 8), (4, 5, 3, 6)}

We obtain the following terms of the reliability expression:

(i) <u>For Path set # 1</u>

$p_1 p_6$

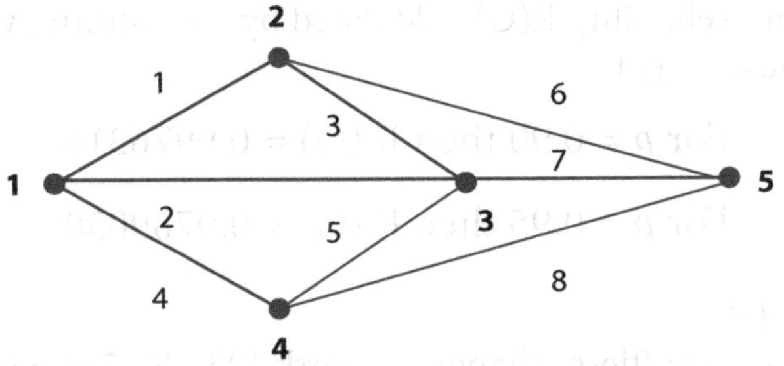

Figure 4.3 5 nodes and 8 links network for reliability evaluation.

(ii) For Path set # 2

$$(1 - p_1p_6)\, p_2p_7$$

(iii) For Path set # 3

$$(1 - p_1p_6)\,(1 - p_2p_7)p_4\, p_8$$

(iv) For Path set # 4

$$(1 - p_2)\,(1 - p_4p_8)(1 - p_6)p_1\, p_3p_7$$

(v) For Path set # 5

$$(1 - p_1)\,(1 - p_4p_8)(1 - p_7)p_2\, p_3p_6$$

(vi) For Path set # 6

$$(1 - p_4)\,(1 - p_6)(1 - p_7)p_2\, p_5p_8$$
$$(1 - p_1)\,(1 - p_3)(1 - p_6)(1 - p_7)p_1p_3p_6p_7$$

(vii) For Path set # 7

$$(1 - p_1)\,(1 - p_2)(1 - p_8)p_4\, p_5p_7$$
$$(1 - p_2)(1 - p_3)\,(1 - p_6)(1 - p_8)(1 - p_6)p_1p_4p_5p_7$$

(viii) For Path set # 8

$$(1 - p_2)\,(1 - p_4)(1 - p_6)(1 - p_7)p_1p_3p_5p_8$$

(ix) For Path set # 9

$$(1 - p_1)\,(1 - p_2)(1 - p_7)(1 - p_8)p_3p_4p_5p_6$$

With the assumption of equal links reliabilities, p, the above terminal reliability expressions would reduce to:

$$R(G) = 3p^2 + 4p^3 - 9p^4 - 10p^5 + 27p^6 - 18p^7 + 4p^8$$

The system reliability R(G) calculated by substituting various values of p. Thus, we get

$$\text{For } p = 0.90 \text{ then } R\,(G) = 0.9976316$$

$$\text{For } p = 0.95 \text{ then } R\,(G) = 0.9789030$$

Illustration 4.4
We consider a shuffle-exchange network [23, 36, 71] with an extra stage (**SEN+**) as shown in Figure 4.4.

This 8×8 SEN+ is generally represented having 8 sources and 8 destinations connected by **16 switches** through **4 stages.**

We obtain the following terms of the reliability expression:

(i) For Path set # 1
$p_1 p_5 p_{10} p_{16}$

(ii) For Path set # 2
$(1 - p_5 p_{10})\, p_1 p_6 p_{12} p_{16}$

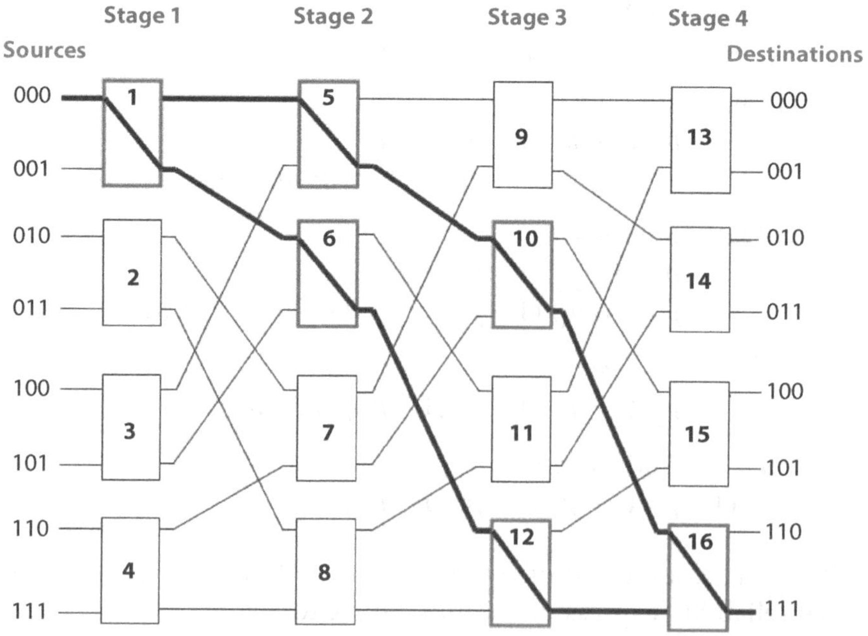

Figure 4.4 8×8 SEN+ showing two disjoint and redundant paths for S-D node pair (0, 7).

With the assumption of equal links reliabilities, p, the above terminal reliability expressions would reduce to:

$$R(G) = 2p^4 - p^6$$

The system reliability R(G) calculated by substituting various values of p. Thus we get

For $p = 0.90$ then R (G) = 0.780759

For $p = 0.95$ then R (G) = 0.893921

Reliability is one of the major considerations and important figures-of-merit in the design of fault tolerant MINs.

Reliability is an essential requirement for improving performance of MIN. Thus reliability evaluation is a vital step in all MIN system studies.

This two-step procedure is used to evaluate MIN reliability in next section.

- Enumeration of minimal path sets of the network, and
- Efficient method to handle these path sets using hybrid multiple variable inversion (MVI) algorithm to compute network reliability expression.

4.1.3 Reliability Evaluation Techniques Comparison

Available literature (Chapter 3) on reliability evaluation of MINs use various techniques like

- Continuous time Markov chains (CTMC) [78]
- Reliability bounds [85–98]
- Reliability block diagram [66–72]
- Decomposition method [84]
- Monte Carlo simulation [99–104]

However, all these methods provide inaccurate reliability values and there is always a demand in finding the exact reliability of multistage interconnection networks for both small and large scale complex parallel interconnection systems.

Aiming to evaluate exact reliability this chapter applied MVI approach and compared the results with reliability bounds [23] and reliability block diagram (RBD) [71] evaluation techniques in one of the most famous MIN called shuffle-exchange network (SEN) and its variants namely SEN+ (SEN with one additional stage) and SEN+2 (SEN with two additional stages).

Reliability bounds and reliability block diagram (RBD) methods are commonly used methods in SEN systems. Shuffle-exchange network and extra-stage shuffle-exchange networks [23, 36, 71] will be examined because they have been widely accepted as a practical interconnection system owing to the size of their SEs and uncomplicated configuration.

Shuffle-exchange network (SEN) is one of the most popular MINs. There are two main reasons for the attractiveness of this network: low hardware cost and efficient routing mechanism (i.e. self-routing). However, the network is a typical MIN that provides a single path between each source-destination pair.

Three types of SENs are studied: SEN, SEN with an additional stage (SEN+), and SEN with two additional stages (SEN+2). Measures of network system reliability: terminal and broadcast reliability are evaluated to study the influence of additional stages on SEN reliability. SEN with three stages, 12 switches (SEs), and 32 links are shown in Figure 2.5 and 3.2. SEN+ shown in Figure 2.5 and Figure 4.4. SEN+2 are shown in Figure 2.17. Detailed switch structure of SEN, SEN+ and SEN+2 is given in section 2.2.

Reliability evaluation is a vital step in all system reliability studies. Based on the literature, terminal pair reliability of SEN systems is calculated by Gunawan [23] using reliability bounds method and by Bistouni [71] using reliability block diagram approach.

We propose to calculate SEN reliability by enumerating minimal path sets and then evaluate reliability expression in compact form of sum of disjoint products (SDP) using improved multi-variable inversion (MVI) algorithm [124].

4.1.3.1 Terminal Reliability of SEN, SEN+ and SEN+2

Terminal reliability is defined as the probability of successful communication between an input and an output switch. In this section, terminal reliability of SEN, SEN+, and SEN+2 is compared.

For SEN and SEN+, same TR values are obtained for all three methods (reliability bounds, reliability block diagram and sum of disjoint products). These values are shown in Table 4.5. But the evaluated reliability values for SEN+2 vary. MVI approach provides exact reliability; RBD method provides lower values but very near to exact values and shown in Table 4.6.

Terminal reliability values for the three networks (SEN, SEN+ and SEN+2) of size 8×8 are calculated. These results indicate that SEN+2 operate slightly better than SEN+, however, it is much better than SEN. SEN+ and SEN+2 continuously outperform SEN and they are not alike in terms of reliability. SEN+2 is more efficient than other two networks.

4.1.3.2 Broadcast Reliability of SEN, SEN +, and SEN+2

Broadcast reliability defined as the probability of successful communication between the source and all network destinations. Enumerating the entire path sets for the network which gives four (4) paths for SEN+ and twelve paths (12) for SEN+2. The calculated broadcast reliability values for SEN by all three methods are same and compared in Table 4.7.

Broadcast reliability results of SEN+ and SEN+2 is given in Tables 4.8–4.9. SEN has the weakest results and SEN+2 values are dominant. SEN+2 achieve a higher broadcast reliability than the other two networks.

Table 4.5 Terminal reliability of 8×8SEN and SEN+.

Switching reliability	Terminal reliability of SEN	Terminal reliability of SEN+
0.99	**0.970299**	0.979712
0.98	**0.941192**	0.958894
0.96	**0.884736**	0.915935
0.95	**0.857375**	0.893921
0.90	**0.729000**	0.780759

Table 4.6 Terminal reliability of 8×8 SEN+2.

Switching reliability	R_n (SEN+2) (MVI method)	R_n (SEN+2) (RBD)	R_n (SEN+2) (Bounds)
0.99	**0.9798963**	0.979708	0.924345
0.98	**0.9595733**	0.958865	0.856787
0.96	**0.9182251**	0.915721	0.742687
0.95	**0.8971944**	0.893519	0.694677
0.90	**0.7888415**	0.778213	0.520995

Overall, we conclude that although SEN+2 achieves a higher broadcast reliability than the other two networks, adding two stages to SEN is more efficient than adding one stage, because reliability can be increased considerably by adding two stages to SEN, while it is increased only slightly by adding one stage.

4.1.3.3 Comparison

Reliability of three networks, SEN, SEN+, and SEN+2 is analyzed from two perspectives, terminal and broadcast in order to investigate the impact of increasing the number of switching stages on the reliability of MINs.

The purpose of this study has been to derive reliability values of shuffle exchange network (SEN) systems and to compare with other reliability evaluation techniques. A reliability analysis of the SEN,

Table 4.7 Broadcast reliability of 8×8 SEN [Three Methods].

Switching reliability	R_n (SEN)
0.99	**0.932065**
0.98	**0.868126**
0.96	**0.751447**
0.95	**0.698337**
0.90	**0.478297**

Table 4.8 Broadcast reliability of 8×8 SEN+.

Switching reliability	Rn (SEN+) (MVI method)	Rn (SEN+) (RBD)	Rn (SEN+) (Bounds)
0.99	**0.9503338**	0.950151	0.950334
0.98	**0.9014617**	0.900795	0.901462
0.96	**0.8067559**	0.804540	0.806756
0.95	**0.7611920**	0.758041	0.761192
0.90	**0.5548722**	0.547124	0.554872

Table 4.9 Broadcast reliability of 8×8SEN+2.

Switching reliability	Rn (SEN+2) (MVI Method)	Rn (SEN+2) (RBD)	Rn (SEN+2) (Bounds)
0.99	**0.95069187**	0.950515	0.897863
0.98	**0.90274140**	0.902115	0.809831
0.96	**0.81082116**	0.808875	0.667692
0.95	**0.76683668**	0.764166	0.610132
0.90	**0.56699802**	0.561638	0.408837

SEN+, and SEN+2 was conducted in [23] to assess the impact of several additional stages on the SEN reliability. In this study, it is concluded that the SEN+ has the highest reliability and SEN+2 has the lowest one. SEN achieves higher reliability than SEN+2.

Therefore, a re-analysis of the reliability of three networks of SEN, SEN+, and SEN+2 was performed in [71]. The results of this study indicated that SEN+ and SEN+2 have the same functionality in terms of reliability. Also, the reliability of these networks is always more than the single-path SEN reliability. Therefore, adding one switching stage to SEN is preferable to adding two switching stages.

Finally, we have performed the study and our results demonstrated that SEN+2 always perform better than SEN+ and also the reliability of these two networks is always more than the single-path SEN. The reliability analysis is much more accurate than other previous analysis as we have used exact reliability evaluation methods.

Thus, for a simple 8×8 SEN systems evaluated reliability values varies with earlier approaches increasing uncertainty and affecting suitability of designs for large scale parallel systems. Therefore, it is necessary to evaluate exact reliability in modern system design.

4.2 Reliability Analysis of Multistage Interconnection Networks

Reliability of a MIN [125] is defined as the probability that a MIN will be able to provide the required interconnection for a specified period of time under a given set of operating conditions. Numerous methods had been devised to enumerate MIN reliability and discussed in Chapter 3.

The reliability and total number of redundant and disjoint paths for every Multistage interconnection network are calculated by enumerating minimal path sets and then evaluating reliability expression in compact form for sum of disjoint products (SDP) using multi-variable inversion (MVI) method. Present work has attempted to evaluate and compare various recent and industrial used MINs topologies based on terminal reliability for multistage interconnection networks available from literature and shown in Table 4.13.

Case 1: Omega Network

8×8 Omega network [35] is generally represented by connecting 8 Sources and 8 Destinations through 12 switches and 32 links shown in Figure 4.5.

The network is represented or fed to a computer program through its adjacency matrix. Adjacency matrix is 'N×N' matrix, whose N rows and N columns corresponds to N nodes (switches in MIN) of the network.

Table 4.10 shows the adjacency matrix for the network in Figure 4.5. Any row having '1' in any column of the adjacency matrix indicates that two nodes are interconnected by a link. For example, 5th row of

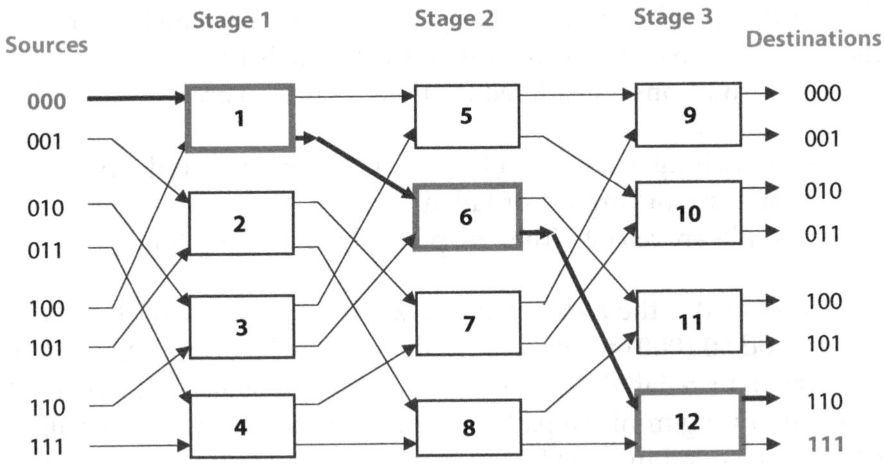

Figure 4.5 8×8 omega network.

Table 4.10 Adjacency matrix for the network in Figure 4.5.

Nodes/ switches	1	2	3	4	5	6	7	8	9	10	11	12
1	0	0	0	0	1	1	0	0	0	0	0	0
2	0	0	0	0	0	0	1	1	0	0	0	0
3	0	0	0	0	1	1	0	0	0	0	0	0
4	0	0	0	0	0	0	1	1	0	0	0	0
5	0	0	0	0	0	0	0	0	1	1	0	0
6	0	0	0	0	0	0	0	0	0	0	1	1
7	0	0	0	0	0	0	0	0	1	1	0	0
8	0	0	0	0	0	0	0	0	0	0	1	1
9	0	0	0	0	0	0	0	0	0	0	0	0
10	0	0	0	0	0	0	0	0	0	0	0	0
11	0	0	0	0	0	0	0	0	0	0	0	0
12	0	0	0	0	0	0	0	0	0	0	0	0

Table 4.10 has value '1' in the columns named 9, 10 which interprets that the switch 5 connects to switch 9 and switch 10.

The assumptions considered for this evaluation are:

- Switching elements (1, 2 ....12) can have only two states; working = 1 or failed = 0
- Links are reliable. They are perfectly in working condition

Now consider the 8×8 omega network for the flow of data from source node 0 (000) wants to send data to destination node 7 (111). Terminal pair reliability evaluation of omega networks is carried out by enumerating minimal path sets through the algorithm. The algorithms are coded in MATLAB 9.0 and implemented on a computer with a 3.00 GHz Intel (R) Core (TM) 2 Duo processor and 2 GB of RAM with Windows 7 operating system. We get,

Number of Redundant Paths for any source-destination pair = 1

Number of disjoint Paths for any source-destination pair = 1

Enumerated minimal path set: 1-6-12

Terminal reliability expression is evaluated from these minimal paths set and calculated using MVI method. System reliability can be computed by providing assumed different switch reliability values to the reliability expression.

Terminal pair reliability expression for Omega network = r^3

System Reliability = **0.7290** (When r = 0.90)

Case 2: Gamma Networks

The gamma interconnection network [41] is discussed in Section 2.2.1 and shown in Figure 2.14. We can deduce that path enumeration in these networks provide various paths for different tag value and reliability expression will also vary for every tag value and shown in Table 4.11.

Table 4.11 Terminal reliability expression of 8×8 gamma networks.

Tag value (δ)	Number of paths	Terminal reliability
0	1	r^4
1	4	$3r^4 - 3r^5 + r^6$
2	3	$2r^4 - r^5$
3	5	$3r^4 - 3r^5 + r^6$
4	2	r^4
5	5	$3r^4 - 3r^5 + r^6$
6	3	$2r^4 - r^5$
7	4	$3r^4 - 3r^5 + r^6$

Case 3: CGIN

Cyclic Gamma Interconnection Network (CGIN) [46] is a revised GIN [41]. CGIN is discussed in section 2.2.1 and shown in Figure 2.23. In this network we are getting number of alternate paths as 2, 4 and 5 and shown in Table 4.12.

Table 4.12 Terminal reliability expression of $8\times8CGIN^0_3$.

Tag value (δ)	Number of paths	Terminal reliability
0	5	$5r^4 - 4r^5 - 3r^6 + 4\,r^7 - r^8$
1	4	$4r^4 - 2r^5 - 4r^6 + 4\,r^7 - r^8$
2	4	$4r^4 - 2r^5 - 3r^6 + 2r^7$
3	2	$2r^4 - r^6$
4	2	$2r^4 - r^6$
5	2	$3r^4 - 3r^5 + r^6$
6	4	$4r^4 - 2r^5 - 3r^6 + 2r^7$
7	4	$4r^4 - 2r^5 - 4r^6 + 4\,r^7 - r_8$

Terminal pair reliability (TPR) is the most commonly used reliability performance index of MINs. Then we extend our work to trace the minimal path sets of various existing topologies of MINs and terminal pair reliabilities are evaluated.

Previous works has been limited only to computing reliability of more general MINs like Omega network [35], Shuffle exchange network [36] and Gamma network [46]. This is the very first attempt to calculate and compare MIN reliability of various topologies.

Terminal reliability for different MINs are computed by generating all possible paths for each source and destination pair of the network and then evaluating terminal reliability from the minimal paths assuming different switch reliability values. The terminal reliability values, evaluated for various MINs are shown in Table 4.13.

MINs are tabulated with number of redundant and disjoint paths values. Scalable crossbar [66] is a non-blocking network with a very low-cost layout outperforms other single tolerant networks and traditional MINs ([31–39, 41] and [44]).

Terminal reliability of scalable crossbar network (SCN) [66] is higher for lower switch (node) reliability values as shown in Table 4.13. SCN is designed with only one switching element between any source-destination node pair. Therefore, terminal reliability value is always its switch reliability values i.e. RSCN (t) = r (t).

Networks (Banyan [31], baseline, reverse baseline [32], delta [33], cube [34], omega [35], reverse omega [32] and shuffle exchange [36]) are poor fault tolerant because the total number of paths for any source-destination node pairs is only one. Unique-path MINs are poor fault tolerant with very low reliability values using very low-cost layout. Replicated Benes [67] provides two disjoint paths and high reliability values comparing other two fault tolerant networks.

IEGN [68] achieved significant tolerance to faults with relatively low-cost layout than other three disjoint networks ([49] and [69]). IEGN [58] topology is fault tolerant with higher reliability values providing more multiple paths using multiplexers and demultiplexers. It should be noted that, as network size increases, the reliability of the network might be reduced.

From Table 4.13, MINs ([63, 64]) are highly fault tolerant networks with four disjoint paths. GINs [31–50] are slightly fault tolerant and having high reliability values using complex design. Other MINs

Table 4.13 Multistage interconnection networks terminal reliability evaluation.

Network types	Number of paths	Disjoint paths	TPR for switch reliability (r) =				
			0.95	**0.90**	**0.80**	**0.70**	**0.60**
Banyan [31], Baseline, Reverse Baseline, Reverse Omega [32], Delta [33], Cube [34], Omega [35], Shuffle Exchange [36]	1	1	0.8574	0.7290	0.5120	0.3430	0.2160
Butterfly [37]	1	1	0.8145	0.6561	0.4096	0.2401	0.1296
Dot network [38]	1	1	0.7737	0.5904	0.3276	0.1681	0.0778
Gamma [41]	1	1	0.8145	0.6561	0.4096	0.2401	0.1296
Monogamma GIN [44]	2	1	0.8552	0.7217	0.4915	0.3121	0.1815
Theta [39]	3	1	0.8031	0.6223	0.3377	0.1597	0.0646
Scalable crossbar [66]	**1**	**1**	**0.9500**	**0.9000**	**0.8000**	**0.7000**	**0.6000**
Augmented SEN [36]	4	2	0.8845	0.7480	0.4689	0.2472	0.1085
2-ary 3-tree (fat-tree) [37]	16	2	0.8935	0.7782	0.5448	0.3396	0.1848
PHI [40]	5	2	0.8742	0.7197	0.4198	0.2041	0.0822

(Continued)

Table 4.13 Multistage interconnection networks terminal reliability evaluation. (*Continued*)

Network types	Number of paths	Disjoint paths	TPR for switch reliability (r) =				
			0.95	**0.90**	**0.80**	**0.70**	**0.60**
Extra Stage GIN [42]	8	2	0.8997	0.7979	0.5916	0.3950	0.2300
Fully chained GIN [47]	64	2	0.8946	0.7791	0.5399	0.3299	0.1769
Partial chained GIN [47]	18	2	0.8955	0.7844	0.5587	0.3564	0.2006
No Backtracking GIN [48]	4	2	0.8960	0.7873	0.5734	0.3841	0.2332
Minimal links traversed GIN [50]	4	2	0.8960	0.7873	0.5734	0.3841	0.2332
Combining switches MIN [51]	4	2	0.8960	0.7873	0.5734	0.3841	0.2332
Balanced Gamma [52]	4	2	0.8979	0.7938	0.5898	0.4057	0.2540
Benes [57]	4	2	0.8935	0.7782	0.5448	0.3396	0.1848
Extra-Stage Benes [67]	8	2	0.8931	0.7755	0.5323	0.3169	0.1596
Extra Stage SEN (SEN+) [71], Extra Stage Cube [58]	2	2	0.8939	0.7807	0.5570	0.3626	0.2125
Parallel Benes [67]	16	2	0.9799	0.8933	0.7766	0.5331	0.3140

(*Continued*)

Table 4.13 Multistage interconnection networks terminal reliability evaluation. (*Continued*)

Network types	**Number of paths**	**Disjoint paths**	**TPR for switch reliability (r) =**				
			0.95	**0.90**	**0.80**	**0.70**	**0.60**
SEN+2 and Omega +2 [71]	4	2	0.8971	0.7885	0.5658	0.3698	0.1998
Kappa network [56]	4	2	0.8979	0.7939	0.5899	0.4058	0.2540
Replicated Benes [67]	**8**	**2**	**0.9886**	**0.9508**	**0.7928**	**0.5638**	**0.3354**
Extra group network [68]	3	3	0.8967	0.7770	0.5082	0.2750	0.1227
3-disjoint paths GIN [49]	4	3	0.9018	0.8057	0.6246	0.4516	0.2971
Enhanced IADM [49]	7	3	0.9023	0.8082	0.6285	0.4600	0.3087
Pars Network [69]	6	3	0.9006	0.7968	0.5684	0.3447	0.1725
Improved extra group [68]	**12**	**3**	**0.9006**	**0.7971**	**0.5686**	**0.3424**	**0.1675**
Extended Generalized SEN [63]	8	4	0.9025	0.8089	0.6203	0.4173	0.2281
Incomplete GIN [55]	6	2	0.8997	0.7979	0.5916	0.3950	0.2300

(*Continued*)

Table 4.13 Multistage interconnection networks terminal reliability evaluation. (*Continued*)

Network types	**Number of paths**	**Disjoint paths**	**TPR for switch reliability (r) =**				
			0.95	**0.90**	**0.80**	**0.70**	**0.60**
Generalized SEN [60]	2	2	0.8940	0.7807	0.5570	0.3626	0.2125
Augmented Baseline [59]	8	2	0.8933	0.7766	0.5331	0.3140	0.1538
Tandem Omega [62]	32	2	0.8933	0.7766	0.5331	0.3140	0.1538
2- D ring Banyan [61]	>50	3	0.9024	0.8089	0.6312	0.4605	0.2986
Clos Network [64]	**4**	**4**	**0.9025**	**0.8099**	**0.6390**	**0.4860**	**0.3508**
Modified Baseline [65]	1	1	0.8145	0.6561	0.4096	0.2401	0.1296

([37, 52, 60] and [62–69]) are notable high-fault tolerant topologies with some modern design ideas.

Recently developed Clos [64] network is highly fault tolerant network with higher reliability values using modest layout. They outperform other MINs because of complex and costly ring-banyan [61] and augmented baseline [59]. They provide dynamic routing which enhances the fault tolerant capability.

4.3 Summary

This chapter has attempted to evaluate and compare various recent and industrial used MINs topologies based on terminal reliability for MINs available from literature.

In a rational methodology, before deciding whether to improve the reliability of a system, a reliability assessment for each component in the system is essential. This assessment is known as the reliability importance analysis (or sensitivity analysis) to clarify the role of each component impact on entire system reliability. Then, the weaknesses of a system can be revealed and a reliability engineer will be able to choose the critical parts of the system to implement fault tolerance/ fault avoidance techniques.

Thus, available disjoint and redundant paths of various existing MINs are evaluated by a simple and efficient path-set enumeration method along with hybrid multi-variable inversion (MVI) algorithm, giving exact reliability values. Present reliability evaluation paves the way of research focus on developing higher disjoint, reliable and highly fault-tolerant MINs than the existing topologies without increasing its hardware complexity.

5

Comprehensive MIN Reliability Paradigms Evaluation

5.1 Introduction

In recent years' parallel and distributed computing largely depends on computer communication networks. Moreover, it is a well-known fact that every component or system is prone to failure and probability of system failure increases with increase in the system size and complexity. Therefore, it is very essential to compute and ensure the reliability of these networks, growing complex day by day.

Reliability modeling and computation is necessary for analyzing these networks and improve the weak links using reliability and design techniques. Considering the present developments in communication hardware industry, where the network components are becoming more reliable and non-repairable and reliability becomes an important performance measure for computer communication networks.

Switch/link failures in Interconnection Networks lead to dilapidation of network interconnections which ends in unbalanced traffic load. Therefore, it is essential to achieve high fault-tolerance through reliable interconnects. A reliable interconnect [66–73] is a promising solution in building a fault-tolerant system to provide successful transmission.

A network designer has to choose a suitable fault tolerant and highly reliable MIN architecture to avoid performance degradation and to route multiple communication tasks concurrently. To ensure desired performance of the networks, the reliability evaluation of

Neeraj Kumar Goyal and S. Rajkumar. Interconnection Network Reliability Evaluation: Multistage Layouts, (115–156) © 2020 Scrivener Publishing LLC

MIN is quite evident. The reliability evaluation [85–98] of MIN can be sub-divided into following different cases:

- Two terminal or terminal pair reliability (TPR): To send messages from a source node s to a terminal node t. The terminal pair reliability of a network is defined as the probability of having at least one operational path between the nodes s and t.
- Global or all terminal reliability (ATR): Another communication operation is complete network connectivity. The all terminal reliability of a network is defined as the probability that for every node pair there exist an operational path to connect them; or equivalently, the probability that there exists a working spanning tree.
- Broadcast reliability: The probability that a single source node is able to broadcast data to all terminal nodes.

According to the reported researches [19, 66–72] and literatures, there is very limited reliability comparison of MIN architectures. Some literatures calculated two-terminal network reliability (TTNR) commonly known as the source-terminal reliability for selected MINs as discussed in Chapter 3. Other researchers extend to broadcast reliability and all-terminal network reliability (ATNR).

Reliability studies on MINs [79–80, 99–104] are discussed in Chapters 3 and 4. These evaluated only traditional reliability measures (two/broadcast/all terminal reliability). With increasing number of input and output nodes MINs, reliability evaluation of multi-cast nodes is mandatory. At the onset of the fourth generation of communication, the communication industries are in urge of reliable communication networks. To harmonize with users' expectations, a network needs to establish communications between different nodes of the network.

In other words, a network has to transmit a signal or commodity between different pairs of nodes of a network simultaneously. Communication between multiple-sources multiple-destinations offers several advantages like increased performance, improved reliability and decreased costs through resource sharing. So, we have extended the traditional reliability evaluation to include a new reliability measure as multi- source multi-terminal reliability (MSMT).

- **Multi Source - Multi Terminal Reliability** (MSMTR): The probability of successful communication between given set of multiple source nodes (S) and multiple terminal nodes (T) in the network. The probability that a network can simultaneously transmit required amount of flows between various specified node-pairs of a network for a given period of time under defined environmental conditions.

Available literature (Chapter 3) on reliability evaluation of MINs use various techniques like

- Continuous time Markov chains - Network size increases, they are arduous and complex.
- Reliability bounds
- Reliability block diagram

Reliability bounds and RBD: As network size and complexity increases, reliability evaluation takes long time. Therefore, it becomes infeasible to attain accurate results.

- Decomposition method
- Monte Carlo simulation - Several runs are used to have a better approximation value.

However, there is always a demand in finding the exact reliability of multistage interconnection networks for both small and large scale complex parallel interconnection systems.

Individual works presented in literature in general, calculated one of the reliability parameters (TPR/ATR/KTR) for the common MIN like omega network [35], shuffle exchange network [36] and cube network. All terminal reliability parameters were evaluated for complicated switching topology like gamma network [41], balanced GIN [52] and ring banyan network.

Reliability of major recently developed MINs [55, 61–63, 65] are yet to be evaluated. Previous works has been limited only to computing more general and traditional measures of reliability. There is lack of literature investigating all the reliability parameters for all MINs. It clearly motivates us for a comprehensive reliability analysis and comparison.

Aiming to evaluate exact reliability this chapter applied simple and computationally efficient approach given by Goyal [125]. This approach is based on enumerating minimal path sets of a network. This approach is easily programmable and it is illustrated with an example, demonstrating the procedure and its effectiveness. Network reliability computation using these path sets is carried out using hybrid multi-variable inversion (MVI) algorithm given by Chaturvedi and Misra [124].

Multi source Multi terminal reliability evaluation can be divided in following two major steps:

- A simple and fast method of enumerating all path sets [125].
- Reliability expressions from these paths set are evaluated by using hybrid multiple variable inversion (MVI) algorithm [124].

The major contributions of this chapter are summarized as follows:

- Efficient approach for enumerating path sets and evaluates exact reliability of MIN. The method is supported by well explained examples at each step.
- A comprehensive evaluation of different reliability measures (All-Terminal and broadcast reliability) is also presented.
- A new reliability paradigm is introduced. In context of MINs, multi-source multi-terminal reliability is defined as the probability that a combination of terminal nodes is connected to a combination of source nodes for given period of time under given use conditions.
- Then we extended our work on various benchmark MINs to trace the minimal path sets of various MINs and different reliability paradigms are evaluated and compared.

5.2 Reliability Evaluation Approach

A Multistage interconnection network is modeled by a probabilistic connected graph. Multistage interconnection networks (MINs) connect N number of input nodes to N number of output nodes through various sizes and various stages of intermediate switching elements. MINs are stage wise networks. MIN has structure in which connections exist to same layer or adjacent layers only. Connections between non-adjacent layers do not exist.

Path set and cut set techniques (POC) [113–124] are the most widely used techniques for terminal pair reliability evaluation of Interconnection networks as these techniques give a compact reliability expression in sum of disjoint product (SDP) form. These techniques first enumerate minimal path (or cut) set for the network and then evaluate reliability (or unreliability) expressions in the SDP form.

Since enumeration of cut set also requires knowledge of basic paths of the network, the path set enumeration becomes a necessity for these techniques. A minimal path set [106–120] is generally used in the sense of a minimal chain of elements required to establish continuity between the two specified terminals (nodes or vertices).

The network reliability can also be computed indirectly taking advantages of duality in graphs [81–83]. In other words, instead of computing reliability one can compute unreliability in the sense of a minimal set of elements whose failure ensures network failure. This set is known as minimal cut set of the network. Therefore, success of all the components in at least one minimal path set ensures network success whereas failure of all components in at least one minimal cut set ensures network failure.

The POC approaches determine reliability (unreliability) expression in the disjoint set form. These approaches [117–124] can be classified as: (1) approaches which give disjoint terms in SVI (single variable inversion) form and (2) approaches which give disjoint terms in MVI (multiple variable inversion). SVI approaches use a simple data presentation that makes these approaches easier for storing and processing the data. The MVI approaches have advantage over SVI of giving more compact reliability expression. Studies on both the SVI and MVI techniques and their reliability evaluation are explained in Chapter 4.

We applied an efficient method for enumerating paths as these path sets are used to obtain system reliability expression in a compact form. Aiming to enumerate minimal path set of a network, this chapter explains a simple and computationally efficient approach.

Several methods are available in literature for the enumeration of path sets as discussed in Chapters 3 and 4. But the major advantage of the Goyal [125] approach is that it is applicable to both simple and complex MIN networks. It is applicable to both unidirectional and bi-directional branches and easily programmable. This approach is illustrated with an example. To make network reliability computation using path sets competitively attractive, MIN reliability (MSMT) carried out in three steps:

- Path Tracing Algorithm (PTA) which enumerates path set of a system in a described format.
- Path Retrieval Algorithm (PRA) retrieves paths from the formatted output of PTA.
- Reliability expressions from these paths set are evaluated by using hybrid multiple variable inversion (MVI) algorithm.

5.2.1 Path Set Enumeration

Hereafter this section emphasizes on the path-based approach for reliability evaluation of MINs. These simple approaches proposed by [Goyal (125)] have favored us to devise methods to look for path sets between any/all specified pair of nodes of a given network viz., source and sink. The technique also enumerates path sets for broadcast and all-terminal path sets in an efficient manner. They also enumerate both redundant and disjoint paths for any/all S-D node pair. Two stepwise algorithms namely the PTA and PRA are presented in this section.

5.2.1.1 Assumptions

The assumptions considered for this study are:

- A Switching elements (SEs) can only have two states; working = 1 or failed = 0
- All SE failures are random and statistically independent.
- SE is assumed to be less reliable than the link and cannot be repaired. Therefore, links are assumed to be perfectly reliable (reliability = 1)

- All SEs have identical reliability.
- All SEs in the first and last stages are assumed to be working.

5.2.1.2 *Applied Approach*

A major problem with computation of reliability measures for a large and complex network, with desired level of accuracy, is the required computation time and complexity of analysis. It is difficult to compute reliability of real-life MIN networks, which are quite large, using desktop computers due to long execution time and high memory requirements. Such computations are usually performed on high-end processors for critical systems.

An extensive research has been carried out to minimize the efforts required for these computations and develop useful tools for network designers. Goyal [125] described an approach to determine minimal-path-sets (or minimal-tie-sets) of complex networks. It is a very versatile approach and easily amenable to computer implementation. It evaluates minimal path sets in two steps.

The first step, named as "Path Tracing Algorithm (PTA)", enumerates minimal path sets of a network graph in a described format. Second step path retrieval algorithm (PRA) retrieves paths from the formatted output of PTA. It claims applicability of the algorithm to both simple and complex networks comprised of directional and/or bi-directional branches. Two stepwise algorithms namely the PTA and PRA are presented in this section. Application of these algorithms, in sequence, gives minimal path sets of a network.

Various steps involved in this algorithm can be stated as follows:

1. First the algorithm starts with building the Connection Matrix.
2. Source and sink nodes are noted.
3. It begins with the execution of PTA in a formatted sequence of variable arrays.
4. It forms the PTA intermediate connection matrices.
5. An execution sequence for recursive calls and returned calls of procedure PTA is done.
6. This continues until the sink node is met and then branches are back traced.

7. Complete execution of PTA is done by enumerating all the branches to the sink node.
8. After complete execution of PTA, PRA is called to retrieve paths from the stored formatted data.
9. PRA reads min-tie-set of the system from the formatted data and traced sequence of branches for all retrieved paths.

5.2.1.3 Path Tracing Algorithm (PTA)

The variables to recursive function PTA are initialized as follows:

1. Make the **Previous node** equal to zero as branching has not started yet.
2. Initialize **Counter** to 1.
3. Initialize **NP** to 0.
4. Make the source node row as **Current row.**

Inputs are: CM (connection matrix), Previous node, Counter, and Current row.
Global variables are: Component, Branch number, Counter, Parent branch, End Branch, and NP (number of paths)

1. Set **Start counter** equal to **Counter** value.
2. Read the columns of **Current row**, one by one, as **new element.**
3. Leave out the **diagonal element.**
4. If **new element** is zero, ignore it.
5. Store the other elements.
6. Increment the counter once for each element stored.
7. **Branch number** of the element read will be the current value of the **Counter.**
8. The **Parent branch** of each new element will be equal to the number stored in the **previous node.**
9. The **Node number** of each element is the column number of the element read.

Repeating above listed sequence (step 2 to step 9), all necessary components are read.

10. Modify connection matrix **CM** by changing all entries in column, corresponding to **Current row**, to zero.
11. Set **End counter** equal to **Counter - 1**.
12. Initialize j = **Start counter**.
13. If j is greater than **End counter**, then RETURN from the current recursive procedure **PTA**.
    Otherwise, read **Node number** (j).
14. If **Node number** (j) is output node then store value of j in **End Branch**, increase **NP** and j by 1, and go to *step* 13.

Otherwise, make a new recursive call to procedure **PTA** assigning new input variables as:

- **Previous node** = **Branch number** of the element.
- **Current row** = **Node number** of the element.
- **CM** = **CM**

Then increment j by 1 and repeat *step* 13.

When all the calls from recursive procedure are returned, it means that all minimal paths have been traced. Then go to PRA for path retrieval from the output of PTA.

5.2.1.4 Path Retrieval Algorithm (PRA)

The input variables to PRA are the global variables resulted from PTA. These are Component, Branch number, Parent branch, End Branch, and NP (number of paths).

1. Set index variable i = 1.
2. If i > **NP**, then *STOP*. All paths are retrieved.
3. Read component **End Branch** (i) from stored components.
4. If **Component** is equal to 1, then do not store component name. It only indicates a short.

5. Else, store the element.
6. If **Branch Number**, corresponding to the element, is equal to 0, then increase i by 1 and go to step 2.
7. Else, read element for branch number equal to **Parent Branch** number and go to step 4.

All paths are retrieved from the stored format.

Illustration 5.1

Algorithms PTA and PRA are applied to determine minimal-tie-sets for node pair for the network shown in Figure 5.1 for directed network. The source has assigned the number 1 while the destination as number 5. The connection matrix (CM) presentation is shown in Table 5.1.

Execution results of PTA are shown in Table 5.2 in a formatted sequence of variable arrays. The intermediate connection matrices used in PTA are displayed in Table 5.3. An execution sequence for recursive calls and returned calls of procedure PTA is given in Table 5.4.

In this table, Ret corresponds to call returned from PTA and New corresponds to new recursive call made to PTA. Working of PTA is explained by referring Table 5.2, Table 5.3 and Table 5.4. Initially input variables to PTA are (refer to row #1 in Table 5.4):

- **CM** as shown in Table 5.1; **Previous Node** = 0; **Current Row** =1.

Execution of PTA, up to 10 steps, adds branches from #1 to #3 shown in Table 5.2. These are the elements present in *Current Row*

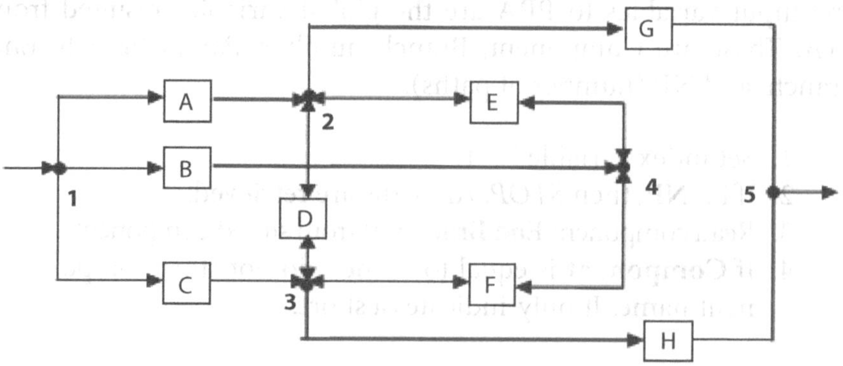

Figure 5.1 Sample network.

Table 5.1 Connection matrix (CM).

Nodes	1	2	3	4	5
1	1	A	C	B	0
2	0	1	D	E	G
3	0	D	1	F	H
4	0	E	F	1	0
5	0	0	0	0	1

(row #1) of *CM*. Then step 11 modifies the *CM* as shown in Table 5.3(a) by making all elements present in column corresponding to *Current Row* equal to 0. Then execution of step 12 and 13 set *Counter* for execution of step 14 and 15 for all added branches (#1 to #3). First added branch is #1 and it doesn't terminate to *sink node* therefore a recursive call to PTA is made with inputs as below (refer to row #2 in Table 5.4)

- **CM** as shown in Table 5.3(a); **Previous Node** = 1; **Current Row** = 2.

First 10 steps execution of PTA result in adding components present in row #2 of Table 5.3(a). It results in addition of branches from #4 to #6 and modification of *CM* as shown in Table 5.3(b) by changing all elements present in column #2 (corresponding to *Current Row*) of Table 5.3(a) to zero. Added branch #4 doesn't terminate to sink node therefore another recursive call to MPTA is made with inputs as below (refer to row #3 in Table 5.4)

- **CM** as shown in Table 5.3(b); **Previous Node** = 4; **Current Row** = 3.

This recursive call results in addition of two more branches #7 and #8 and *CM* is modified as shown in Table 5.3(c) by changing elements present in column #3 to 0. Added branch #7 leads to another call of PTA with inputs as below (refer to row #4 in Table 5.4)

- **CM** as shown in Table 5.3(c); **Previous Node** = 7; **Current Row** = 4.

Table 5.2 Complete Tie-set in formatted form after application of PTA.

Branch Number	**1**	**2**	**3**	**4**	**5**	**6**	**7**	**8**	**9**	**10**	**11**	**12**	**13**
Parent Branch	0	0	0	1	1	**1**	4	**4**	5	**9**	2	2	**2**
Component	A	C	B	D	E	**G**	F	**H**	F	**H**	D	F	**H**
Node Number	2	3	4	3	4	**5**	4	**5**	3	**5**	2	4	**5**
Branch Number	**14**	**15**	**16**	**17**	**18**	**19**	**20**	**21**	**22**	**23**	**24**	**25**	
Parent Branch	11	**11**	12	**16**	3	3	18	**18**	**20**	19	**19**	**23**	
Component	E	**G**	E	**G**	E	F	D	**G**	**H**	D	**H**	**G**	
Node Number	4	**5**	2	**5**	2	3	3	**5**	**5**	2	**5**	**5**	

Table 5.3 Intermediate matrices in PTA.

(a)

#	1	2	3	4	5
1	0	A	C	B	0
2	0	1	D	E	G
3	0	D	1	F	H
4	0	E	F	1	0
5	0	0	0	0	1

(b)

#	1	2	3	4	5
1	0	0	C	B	0
2	0	0	D	E	G
3	0	0	1	F	H
4	0	0	F	1	0
5	0	0	0	0	1

(c)

#	1	2	3	4	5
1	0	0	0	B	0
2	0	0	0	E	G
3	0	0	0	F	H
4	0	0	0	1	0
5	0	0	0	0	1

(d)

#	1	2	3	4	5
1	0	0	C	0	0
2	0	0	D	0	G
3	0	0	1	0	H
4	0	0	F	0	0
5	0	0	0	0	1

(e)

#	1	2	3	4	5
1	0	A	0	B	0
2	0	1	0	E	G
3	0	D	0	F	H
4	0	E	0	1	0
5	0	0	0	0	1

(f)

#	1	2	3	4	5
1	0	0	0	B	0
2	0	0	0	E	G
3	0	0	0	F	H
4	0	0	0	1	0
5	0	0	0	0	1

(g)

#	1	2	3	4	5
1	0	A	0	B	0
2	0	1	0	E	G
3	0	D	0	F	H
4	0	E	0	1	0
5	0	0	0	0	1

(h)

#	1	2	3	4	5
1	0	A	0	0	0
2	0	1	0	0	G
3	0	D	0	0	H
4	0	E	0	0	0
5	0	0	0	0	1

(i)

#	1	2	3	4	5
1	0	A	C	0	0
2	0	1	D	0	G
3	0	D	1	0	H
4	0	E	F	0	0
5	0	0	0	0	1

(j)

#	1	2	3	4	5
1	0	0	C	0	0
2	0	0	D	0	G
3	0	0	1	0	H
4	0	0	F	0	0
5	0	0	0	0	1

(k)

#	1	2	3	4	5
1	0	A	0	0	0
2	0	1	0	0	G
3	0	D	0	0	H
4	0	E	0	0	0
5	0	0	0	0	1

(l)

#	1	2	3	4	5
1	0	0	0	0	0
2	0	0	0	0	G
3	0	0	0	0	H
4	0	0	0	0	0
5	0	0	0	0	1

Table 5.4 Complete execution flow of PTA.

#	Recursive call	Inputs to recursive call			Added branches	Added end branches
		CM(Table)	**Parent Branch**	**Current Row**		
1	New	Table 5.1	0	1	{1,2,3}	–
2	New	Table 5.3(a)	1	2	{4,5,6}	–
3	New	Table 5.3(b)	4	3	{7,8}	–
4	New	Table 5.3(c)	7	4	–	–
5	Ret, Ret, New	Table 5.3(b)	5	4	{9}	{8}
6	New	Table 5.3(d)	9	3	{10}	–
7	Ret, Ret, Ret, New	Table 5.3(a)	2	3	{11,12,13}	{10,6}
8	New	Table 5.3(e)	11	2	{14,15}	–
9	New	Table 5.3(f)	14	4	–	–
10	Ret, Ret, New	Table 5.3(g)	12	4	{16}	{15}
11	New	Table 5.3(h)	16	2	{17}	–
12	Ret, Ret, Ret, New	Table 5.3(a)	3	4	(18,19}	{17,13}
13	New	Table 5.3(i)	18	2	{20,21}	–
14	New	Table 5.3(j)	20	3	{22}	–
15	Ret, Ret, Ret, New	Table 5.3(k)	19	3	{23,24}	{22,21}
16	New	Table 5.3(l)	23	2	{25}	
17	Ret, Ret, Ret					{25,24}

This call doesn't result in any addition of branches since no element is present in row 4 of Table 5.3(c). Therefore, the call is returned and *CM* of Table 5.3(c) is restored. Then PTA, corresponding to returned call, continues for branch #8. Since branch #8 terminates to sink node, its number is added to *End Branch*, which stores the termination points of the branches. There are no other branches added in this call of the procedure therefore it returns. It results in restoration of *CM* shown in Table 5.3(b). It makes another recursive call corresponding to next branch #5 with inputs as below (refer to row#5 in Table 5.4):

- **CM** as shown in Table 5.3(b); **Previous Node** = 5; **Current Row** = 4.

Row #4 of Table 5.4(b) has one element therefore one branch #9 is added. This branch makes another recursive call with inputs as below (refer to row #6 in Table 5.4):

- **CM** as shown in Table 5.3(d); **Previous Node** = 9; **Current Row** = 3.

Row #3 of Table 5.3(d) adds another branch #10 terminating to sink node. Therefore, branch #10 is added to *End Branch*. Two recursive calls are then returned and *CM* shown in Table 5.3(b) is restored. Last added branch #6 in that call terminates to *sink node*. Therefore #6 is added to *End Branch*, another call is returned and *CM* shown in Table 5.3(a) is restored to find leading branches from branch #2 and branch #3.

Similar to above, rest of the branches, shown in Table 5.2 are determined. After complete execution of PTA, PRA is called to retrieve paths from the stored formatted data.

Path Retrieval Algorithm (PRA) reads min-tie-set of the system from the formatted data stored in Table 5.2. Number of paths in the tie-set is equal to *NP*, which is equal to the number of entries in array *End Branch*. Each entry in *End Branch* is taken as starting point and then it traces back through variable *Previous Node* to find components present in corresponding path.

Table 5.5 shows the tracing sequence of branches for all retrieved paths. It can be observed, from Table 5.5, that all paths have starting

Table 5.5 Paths obtained from Table 5.2 after application of PRA.

Path #	Components in path	Sequence of branch numbers followed
1	HDA	8->4->1
2	HFEA	10->9->5->1
3	GA	6->1
4	GDC	15->11->2
5	GEFC	17->16->12->2
6	HC	13->2
7	HDEB	22->20->18->3
8	GEB	21->18->3
9	GDFB	25->23->19->3
10	HFB	24->19->3

branch corresponding to the entries of array *End Branch,* shown in the last column of Table 5.4. Thus 10 path sets are enumerated for the sample network shown in Figure 5.1.

Illustration 5.2: Terminal Reliability Evaluation:
We consider 4-disjoint paths multistage interconnection network as shown in Figure 5.2 generally represented having 8 sources and 8 destinations connected by 24 switches and 72 links are stored in Connection Matrix or CM.

This network is represented or fed to a computer is its connection matrix. Connection matrix is 'N × N' matrix, whose N rows corresponds to N nodes (switches in MIN) of the network. Table 5.6 shows 24×24 connection matrix for the network shown in Figure 5.2 as there are 24 nodes/switches available in this design.

Any row having '1' in any column of the Connection matrix indicates that two nodes are interconnected by a link. For example, 5th row of Table 5.6 has value '1' in the columns named 13, 15 and 19 which interprets that the node/switch 5 is connected with switch 13, switch 15 and switch 19.

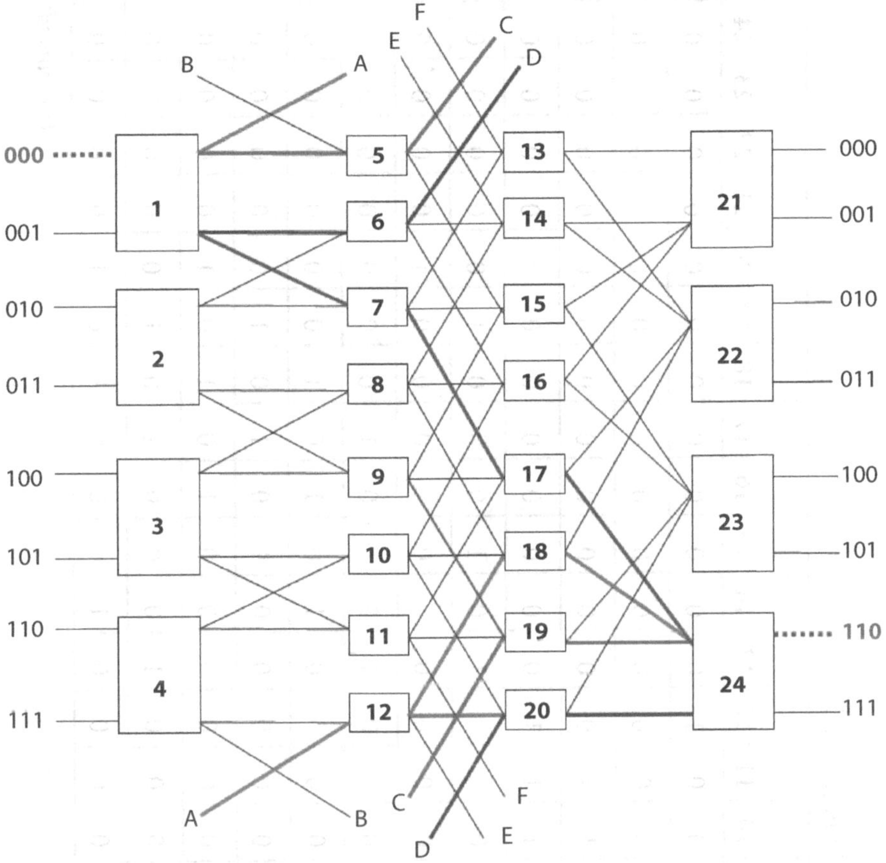

Figure 5.2 MIN showing four disjoints paths and five redundant paths for S-D node pair (0, 6; $\delta = 6$).

Reliability is calculated using this efficient methodology which first enumerates minimal path sets as shown in Table 5.7 and then evaluates reliability expression in compact of sum of disjoint products (SDP) form using multi-variable inversion (MVI) algorithm. Thus in this network,

- Switching elements (1, 2 ....12) can have only two states; working = 1 or failed = 0
- Links are reliable. They are perfectly in working condition
- All SEs in the stage 1 and stage 3 are assumed to be working. Only intermediate switches can fail.

Table 5.6 Connection matrix for the network in Figure 5.2.

Nodes/ switches	1	2	3	4	5	6	7	8	9	10	11	12	13	14	15	16	17	18	19	20	21	22	23	24
1	0	0	0	0	1	1	1	0	0	0	0	1	0	0	0	0	0	0	0	0	0	0	0	0
2	0	0	0	0	0	1	1	1	1	0	0	0	0	0	0	0	0	0	0	0	0	0	0	0
3	0	0	0	0	0	0	0	1	1	1	1	0	0	0	0	0	0	0	0	0	0	0	0	0
4	0	0	0	0	1	0	0	0	0	1	1	1	0	0	0	0	0	0	0	0	0	0	0	0
5	0	0	0	0	0	0	0	0	0	0	0	0	1	0	1	0	0	0	1	0	0	0	0	0
6	0	0	0	0	0	0	0	0	0	0	0	0	0	1	0	1	0	0	0	1	0	0	0	0
7	0	0	0	0	0	0	0	0	0	0	0	0	1	0	1	0	1	0	0	0	0	0	0	0
8	0	0	0	0	0	0	0	0	0	0	0	0	0	1	0	1	0	1	0	0	0	0	0	0
9	0	0	0	0	0	0	0	0	0	0	0	0	0	0	1	0	1	0	1	0	0	0	0	0
10	0	0	0	0	0	0	0	0	0	0	0	0	0	0	0	1	0	1	0	1	0	0	0	0
11	0	0	0	0	0	0	0	0	0	0	0	0	1	0	0	0	1	0	1	0	0	0	0	0
12	0	0	0	0	0	0	0	0	0	0	0	0	0	1	0	0	0	1	0	1	0	0	0	0

(*Continued*)

Table 5.6 Connection matrix for the network in Figure 5.2. *(Continued)*

Nodes/ switches	1	2	3	4	5	6	7	8	9	10	11	12	13	14	15	16	17	18	19	20	21	22	23	24
13	0	0	0	0	0	0	0	0	0	0	0	0	0	0	0	0	0	0	0	0	1	1	0	0
14	0	0	0	0	0	0	0	0	0	0	0	0	0	0	0	0	0	0	0	0	1	1	0	0
15	0	0	0	0	0	0	0	0	0	0	0	0	0	0	0	0	0	0	0	0	1	0	1	0
16	0	0	0	0	0	0	0	0	0	0	0	0	0	0	0	0	0	0	0	0	1	0	1	0
17	0	0	0	0	0	0	0	0	0	0	0	0	0	0	0	0	0	0	0	0	0	1	0	1
18	0	0	0	0	0	0	0	0	0	0	0	0	0	0	0	0	0	0	0	0	0	1	0	1
19	0	0	0	0	0	0	0	0	0	0	0	0	0	0	0	0	0	0	0	0	0	0	1	1
20	0	0	0	0	0	0	0	0	0	0	0	0	0	0	0	0	0	0	0	0	0	0	1	1
21	0	0	0	0	0	0	0	0	0	0	0	0	0	0	0	0	0	0	0	0	0	0	0	0
22	0	0	0	0	0	0	0	0	0	0	0	0	0	0	0	0	0	0	0	0	0	0	0	0
23	0	0	0	0	0	0	0	0	0	0	0	0	0	0	0	0	0	0	0	0	0	0	0	0
24	0	0	0	0	0	0	0	0	0	0	0	0	0	0	0	0	0	0	0	0	0	0	0	0

Table 5.7 Paths obtained from Table 5.2 after application of PRA.

Path #	Sequence of branch numbers followed
1	24->19->5->1
2	24->20->6->1
3	24->17->7->1
4	24->18->12->1
5	24->20->12->1

Illustration 5.3: Broadcast Enumeration:

Table 5.8 Paths obtained after application of PRA.

Path	Sequence of branch numbers followed
1	1-5-13-19-21-22-23-24
2	1-5-6-14-19-21-22-23-24
3	1-5-6-13-20-21-22-23-24
4	1-5-12-13-20-21-22-23-24
5	1-5-12-15-18-21-22-23-24
6	1-5-12-14-19-21-22-23-24
7	1-6-14-20-21-22-23-24
8	1-6-7-16-17-21-22-23-24
9	1-6-7-13-20-21-22-23-24
10	1-6-12-16-18-21-22-23-24
11	1-7-15-17 -21-22-23-24
12	1-7-12-15-18-21-22-23-24
13	1- 7-12-13-20-21-22-23-24
14	1-12-14-20-21-22-23-24

Illustration 5.4: Global or all terminal reliability (ATR) Enumeration:

The similar steps could be followed for other families for MINs to enumerate path sets. The same procedure can be applied to enumerate path sets for broadcast nodes and all terminal nodes. The algorithm can be applied to general network-graph containing undirected, directed or both types of branches.

We consider same 4-disjoint paths gamma interconnection network as discussed in illustration 5.2 for broadcast reliability evaluation. Broadcast reliability evaluation is carried out by enumeration of minimal path sets through the algorithm discussed in section 4.1.12. Various steps involved in the path set enumeration are followed and paths are obtained after application of PRA and shown in Table 5.8. Number of redundant paths for source-destination pair (source 0 to all the destinations) = 14.

We consider same 4-disjoint paths gamma interconnection network as discussed in illustration 5.2 for global reliability evaluation. Various steps involved in the path set enumeration are followed and paths are obtained after application of PRA and shown in Table 5.9. Number of redundant paths for all sources to all destinations nodes is 115.

Illustration 5.5: Multi source - Multi terminal reliability (MS-MTR) Enumeration:

This is the very first initiative attempts which outperforms other literatures to calculate and compare MSMT MIN reliability. They can be defined as "Probability of successful communication between given set of multiple source nodes (S) and multiple terminal nodes (T) in the network".

Benes Network [12] is a rearrangeable non-blocking MIN due to the combination of baseline and reverse baseline network where the intermediate stages are overlapped. Now consider the 8 x 8 Benes Network as shown in Figure 5.3 for the flow of data from multiple sources to multiple destinations.

Table 5.9 Paths obtained after application of PRA for ATR evaluation.

Sequence of Branch Numbers followed (115)		
1 2 3 4 9 12 14 15 17 20 21 22 23 24	1 2 3 4 6 8 11 12 13 16 18 20 21 22 23 24	1 2 3 4 5 6 7 9 10 13 15 18 20 21 22 23 24
1 2 3 4 8 9 12 14 19 20 21 22 23 24	1 2 3 4 5 6 8 10 13 16 18 20 21 22 23 24	1 2 3 4 5 7 8 10 13 14 15 18 19 20 21 22 23 24
1 2 3 4 8 9 12 14 15 18 20 21 22 23 24	1 2 3 4 5 9 12 13 15 17 20 21 22 23 24	1 2 3 4 7 9 10 12 13 14 15 18 19 20 21 22 23 24
1 2 3 4 8 9 12 14 16 17 20 21 22 23 24	1 2 3 4 6 7 11 13 16 17 19 21 22 23 24	1 2 3 4 6 9 11 12 13 15 16 17 18 20 21 22 23 24
1 2 3 4 8 12 14 16 18 20 21 22 23 24	1 2 3 4 6 9 10 12 14 16 18 19 21 22 23 24	1 2 3 4 5 7 8 10 13 14 16 17 19 20 21 22 23 24
1 2 3 4 6 10 14 16 18 20 21 22 23 24	1 2 3 4 7 8 10 12 14 15 17 20 21 22 23 24	1 2 3 4 5 6 8 11 14 15 16 17 20 21 22 23 24
1 2 3 4 6 10 11 13 14 20 21 22 23 24	1 2 3 4 5 7 9 10 13 16 17 19 21 22 23 24	1 2 3 4 6 9 11 12 13 16 17 18 19 21 22 23 24
1 2 3 4 5 9 12 14 15 17 19 21 22 23 24	1 2 3 4 5 6 9 10 13 16 17 20 21 22 23 24	1 2 3 4 7 9 10 12 13 16 18 19 20 21 22 23 24
1 2 3 4 5 8 9 12 14 19 21 22 23 24	1 2 3 4 6 9 10 12 16 17 18 21 22 23 24	1 2 3 4 5 6 8 11 13 15 16 17 18 20 21 22 23 24
1 2 3 4 5 8 12 14 16 18 19 21 22 23 24	1 2 3 4 5 7 9 10 13 15 18 19 21 22 23 24	1 2 3 4 5 7 8 11 12 14 16 17 19 21 22 23 24
1 2 3 4 5 8 9 13 15 18 19 21 22 23 24	1 2 3 4 6 8 11 12 13 16 18 19 21 22 23 24	1 2 3 4 6 9 11 12 13 14 16 18 19 21 22 23 24
1 2 3 4 7 10 11 15 16 17 21 22 23 24	1 2 3 4 7 9 11 12 13 15 17 20 21 22 23 24	1 2 3 4 7 8 9 11 12 13 15 18 20 21 22 23 24
1 2 3 4 6 8 11 12 14 19 20 21 22 23 24	1 2 3 4 5 6 9 10 14 15 18 20 21 22 23 24	1 2 3 4 5 7 8 10 11 14 15 17 20 21 22 23 24
1 2 3 4 6 7 10 11 13 20 21 22 23 24	1 2 3 4 7 9 11 12 13 19 20 21 22 23 24	1 2 3 4 7 8 10 12 14 16 17 20 21 22 23 24

(Continued)

Table 5.9 Paths obtained after application of PRA for ATR evaluation. (*Continued*)

Sequence of Branch Numbers followed (115)		
1 2 3 4 5 7 9 11 13 19 21 22 23 24	1 2 3 4 5 6 9 10 13 15 17 20 21 22 23 24	1 2 3 4 5 6 8 11 14 15 16 17 18 19 21 22 23 24
1 2 3 4 5 8 9 13 16 17 19 21 22 23 24	1 2 3 4 5 6 9 10 13 16 17 19 21 22 23 24	1 2 3 4 5 6 8 9 10 13 15 18 20 21 22 23 24
1 2 3 4 5 8 9 12 15 18 21 22 23 24	1 2 3 4 5 6 9 11 14 15 17 19 21 22 23 24	1 2 3 4 5 6 9 10 12 14 16 17 19 21 22 23 24
1 2 3 4 6 7 10 13 16 18 20 21 22 23 24	1 2 3 4 7 10 11 13 15 17 20 21 22 23 24	1 2 3 4 6 7 9 10 12 13 16 18 19 21 22 23 24
1 2 3 4 6 7 11 13 19 20 21 22 23 24	1 2 3 4 6 8 11 12 14 16 18 19 21 22 23 24	1 2 3 4 5 6 9 11 13 15 17 20 21 22 23 24
1 2 3 4 7 11 13 15 17 19 21 22 23 24	1 2 3 4 5 8 12 13 16 18 20 21 22 23 24	1 2 3 4 6 7 8 11 12 14 16 17 19 21 22 23 24
1 2 3 4 7 10 15 16 17 18 21 22 23 24	1 2 3 4 5 7 8 11 13 15 18 19 21 22 23 24	1 2 3 4 6 9 10 12 14 15 18 20 21 22 23 24
1 2 3 4 6 11 13 14 19 20 21 22 23 24	1 2 3 4 7 8 11 12 13 16 18 20 21 22 23 24	1 2 3 4 6 7 9 11 12 13 16 18 19 21 22 23 24
1 2 3 4 5 7 9 10 15 17 18 21 22 23 24	1 2 3 4 7 8 11 12 13 15 18 19 21 22 23 24	1 2 3 4 6 9 10 12 14 16 17 20 21 22 23 24
1 2 3 4 5 8 13 16 18 19 21 22 23 24	1 2 3 4 7 8 11 12 13 16 17 20 21 22 23 24	1 2 3 4 7 8 11 12 14 16 17 20 21 22 23 24
1 2 3 4 5 6 8 10 13 14 20 21 22 23 24	1 2 3 4 5 6 9 10 14 15 18 19 21 22 23 24	1 2 3 4 5 7 8 10 13 14 15 17 20 21 22 23 24
1 2 3 4 6 8 10 12 16 18 21 22 23 24	1 2 3 4 5 6 9 11 13 16 17 19 21 22 23 24	1 2 3 4 5 6 8 10 11 14 15 17 20 21 22 23 24
1 2 3 4 7 8 10 12 15 16 18 21 22 23 24	1 2 3 4 7 9 11 12 14 15 17 19 21 22 23 24	1 2 3 4 7 9 10 12 13 15 16 18 19 21 22 23 24
1 2 3 4 5 8 12 15 16 18 21 22 23 24	1 2 3 4 7 9 11 12 13 15 18 19 21 22 23 24	1 2 3 4 7 9 10 12 13 14 16 17 19 20 21 22 23 24

(*Continued*)

Table 5.9 Paths obtained after application of PRA for ATR evaluation. (*Continued*)

Sequence of Branch Numbers followed (115)		
1 2 3 4 5 9 13 15 17 19 21 22 23 24	1 2 3 4 5 7 9 10 13 15 17 20 21 22 23 24	1 2 3 4 7 8 10 12 13 16 18 20 21 22 23 24
1 2 3 4 5 9 12 15 17 18 21 22 23 24	1 2 3 4 6 10 11 14 16 17 20 21 22 23 24	1 2 3 4 7 8 11 12 14 15 17 19 21 22 23 24
1 2 3 4 5 7 8 11 15 16 17 21 22 23 24	1 2 3 4 5 7 8 11 14 15 17 19 21 22 23 24	1 2 3 4 6 8 11 12 14 16 17 20 21 22 23 24
1 2 3 4 5 7 9 11 15 17 21 22 23 24	1 2 3 4 5 6 9 11 14 15 17 20 21 22 23 24	1 2 3 4 5 6 8 11 14 15 17 19 20 21 22 23 24
1 2 3 4 6 8 10 12 14 20 21 22 23 24	1 2 3 4 5 6 9 10 14 16 18 19 21 22 23 24	1 2 3 4 6 9 11 12 14 15 16 17 18 19 21 22 23 24
1 2 3 4 6 7 10 16 17 18 21 22 23 24	1 2 3 4 5 8 9 12 13 16 17 20 21 22 23 24	1 2 3 4 5 6 7 8 10 11 14 15 18 19 21 22 23 24
1 2 3 4 5 6 9 11 13 14 19 21 22 23 24	1 2 3 4 5 6 8 10 14 15 18 20 21 22 23 24	1 2 3 4 5 6 9 10 11 14 16 17 19 21 22 23 24
1 2 3 4 6 7 10 11 16 17 21 22 23 24	1 2 3 4 7 8 11 12 14 15 18 19 21 22 23 24	1 2 3 4 5 7 8 10 14 15 17 18 20 21 22 23 24
1 2 3 4 5 8 9 13 14 19 21 22 23 24	1 2 3 4 5 6 8 10 14 16 18 19 21 22 23 24	1 2 3 4 7 8 10 11 12 13 15 18 20 21 22 23 24
	1 2 3 4 5 7 9 10 13 16 18 19 21 22 23 24	1 2 3 4 5 7 8 11 13 16 17 19 21 22 23 24
	1 2 3 4 7 8 10 12 14 15 18 20 21 22 23 24	1 2 3 4 6 7 9 10 11 12 14 15 18 19 21 22 23 24

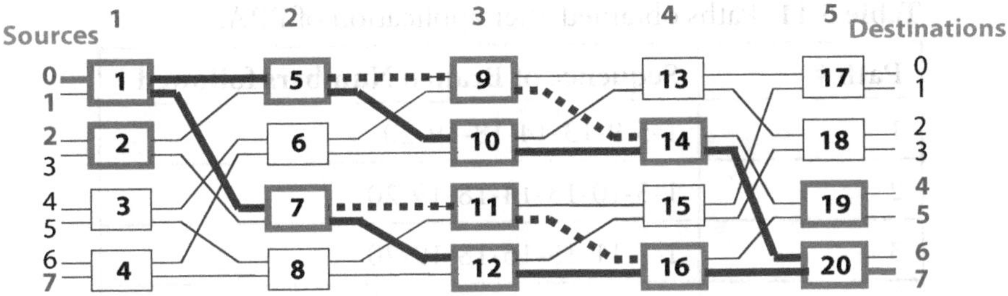

Figure 5.3 8×8 Benes network showing two disjoint paths and four redundant paths for Case 1.

Case 1: 3S/4T
Source nodes = 0, 1, and 2 (Any 3 Sources)
Destination Nodes = 4, 5, 6, and 7 (Any 4 Destinations)
Number of redundant paths for **(3S/4T)** = 4
Enumerated minimal path set:

Table 5.10 Paths obtained after application of PRA.

Path #	Sequence of Branch Numbers followed
1	1 2 5 9 14 19 20
2	1 2 5 10 14 19 20
3	1 2 7 12 16 19 20
4	1 2 7 11 16 19 20

Case 2: 1S/6T
Source node = 0 (Any 1 Source)
Destination nodes = 2, 3, 4, 5, 6, and 7 (Any 6 Destinations)
Number of redundant paths for source-destination pair **(1S/6T)** = 12
Enumerated minimal path set:

Table 5.11 Paths obtained after application of PRA.

Path #	Sequence of Branch Numbers followed
1	1-5-9-13-14-18-19-20
2	1-5-10-13-14-18-19-20
3	1-7-11-15-16-18-19-20
4	1-7-12-15-16-18-19-20
5	1-5-7-9-12-14-15-18-19-20
6	1-5-7-9-11-13-16-18-19-20
7	1-5-7-9-12-13-16-18-19-20
8	1-5-7-9-11-14-15-18-19-20
9	1-5-7-10-11-13-16-18-19-20
10	1-5-7-10-12-13-16-18-19-20
11	1-5-7-10-12-14-15-18-19-20
12	1-5-7-10-11-14-15-18-19-20

Thus 12 path sets for the network shown in Figure 5.3 are enumerated in Table 5.11. Following same fashion, we can enumerate minimal paths for 2-TR, broadcast and global terminal source-destination nodes and also for MINs with multiple-sources and multiple-destinations nodes. We can evaluate the reliability of various MIN topologies in the forthcoming section and shown in Tables 5.12–5.15.

5.3 Reliability Evaluation Using MVI Techniques

Enumeration of minimal path sets of the network is done successfully using Goyal (2006) method. Computation of network reliability expression from these path sets are then handled efficiently by using hybrid multiple variable inversion (MVI) algorithm as discussed in Chapter 4. Let us take a simple example to illustrate the concept to enumerate path sets and the reliability evaluation approach to get disjoint expression using MVI concept.

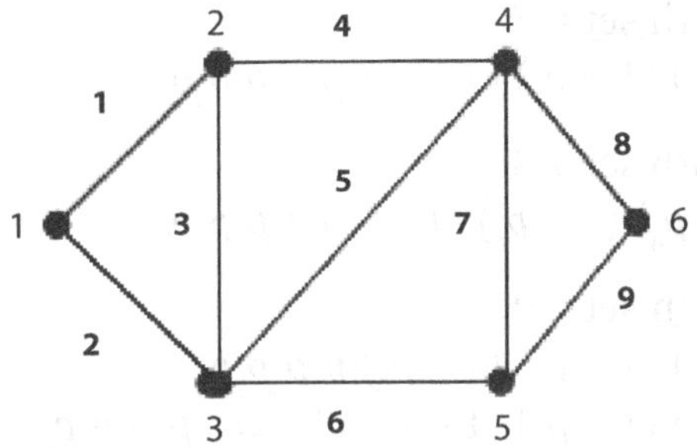

Figure 5.4 8 nodes and 9 links network for reliability evaluation.

Illustration 5.6

The sample network considered is shown in Figure 5.4 contains 6 nodes and 9 links.

The enumeration methods described in Section 5.2.1 enumerates 13 path sets for this network of Figure 4.3 as follows:

Path sets: {**(1, 4, 8)**, **(2, 5, 8)**, **(2, 6, 9)**, **(1, 3, 5, 8)**, **(1, 3, 6, 9)**, **(1, 4, 7, 9)**, **(2, 3, 4, 8)**, **(2, 5, 7, 9)**, **(2, 6, 7, 8)**, **(1, 3, 5, 7, 9)**, **(1, 3, 6, 7, 8)**, **(1, 4, 5, 6, 9)**, **(2, 3, 4, 7, 9)**}

We obtain the following terms of the reliability expression:

(i) For Path set # 1

$p_1p_4p_8$

(ii) For Path set # 2

$(1 - p_1p_4)\, p_2p_5p_8$

(iii) For Path set # 3

$(1 - p_8)\, p_2p_6p_9$

$(1 - p_1p_4)\,(1 - p_5)\, p_2p_6p_8p_9$

(iv) For Path set # 4

$(1 - p_2)\,(1 - p_4)\, p_1p_3p_5p_8$

(v) For Path set # 5

$(1 - p_2)\,(1 - p_8)\, p_1p_3p_6p_9$

$(1 - p_2)\,(1 - p_4)\,(1 - p_8)\, p_1p_3p_6p_8p_9$

(vi) For Path set # 6

$(1 - p_6)\,(1 - p_8)\, p_1p_4p_7p_9$

$(1 - p_2)\,(1 - p_3)\,(1 - p_8)\, p_1p_4p_6p_7p_9$

(vii) For Path set # 7

$(1 - p_1)(1 - p_5)(1 - p_6p_9)\, p_2p_3p_4p_8$

(viii) For Path set # 8

$(1 - p_1p_4)(1 - p_6)(1 - p_8)\, p_2p_5p_7p_9$

(ix) For Path set # 9

$(1 - p_4)(1 - p_5)(1 - p_9)\, p_2p_6p_7p_8$

$(1 - p_1)(1 - p_3)(1 - p_5)(1 - p_9)\, p_2p_6p_7p_8$

(x) For Path set # 10

$(1 - p_2)(1 - p_4)(1 - p_6)(1 - p_8)\, p_1p_3p_5p_7p_9$

(xi) For Path set # 11

$(1 - p_2)(1 - p_4)(1 - p_5)(1 - p_9)\, p_1p_3p_6p_7p_8$

(xii) For Path set # 12

$(1 - p_2)(1 - p_3)(1 - p_7)(1 - p_8)\, p_1p_4p_5p_6p_9$

(xiii) For Path set # 13

$(1 - p_1)(1 - p_5)(1 - p_6)(1 - p_8)\, p_2p_3p_4p_7p_9$

With the assumption of equal links reliabilities, p, the above terminal reliability expressions would reduce to:

$$TR(G) = 3p^3 + 6p^4 - 8p^5 - 21p^6 + 40p^7 - 24p^8 + 5p^9$$

The system reliability R (G) calculated by substituting various values of p. Thus we get

For $p = 0.90$ then R (G) = 0.97718440500

Illustration 5.7

We consider 4-disjoint paths interconnection network as shown in Figure 5.2.

The enumeration methods described in Section 5.2.1 enumerates 13 path sets for this network of Figure 5.2 as follows:

Path sets: {**(1, 5, 19, 24), 1, 6, 20, 24), (1, 7, 17, 24), 1, 12, 18, 24), 1, 12, 20, 24)**}

We obtain the following terms of the reliability expression:

$$TR(G) = 5r^4 - 2r^5 - 7r^6 + 4r^7 + 3r^8 - 2r^9$$

The system reliability R (G) calculated by substituting various values of p. Thus we get

$$\text{For } p = 0.90 \text{ then R (G)} = 0.8091813$$

Reliability evaluation is a vital step in all system reliability studies. In this section all reliability parameters of various MINs are evaluated and compared. MIN reliability is evaluated in three steps:

- Path Tracing Algorithm (PTA) which enumerates path set of a system in a described format
- Path Retrieval Algorithm (PRA) retrieves paths from the formatted output of PTA
- Reliability expressions from these paths set are evaluated by using hybrid multiple variable inversion (MVI) algorithm

The algorithms are coded in MATLAB 9.0 and implemented on a computer with a 3.00 GHz Intel (R) Core (TM) 2 Duo processor and 2 GB of RAM with Windows 7 operating system. All-terminal network reliability (ATNR) and broadcast reliability values are shown in Tables 5.12–5.13. The reliability values are calculated by assuming different switch reliability values. Finally, multi-source multi-terminal reliability (MSMT) evaluations are shown in Tables 5.14–5.15 for node reliability value (r = 0.95 and 0.50).

MINs are tabulated in terms of reliability values in Tables 5.12–5.15. Clos Network [54] and RINs are highly fault-tolerant networks with four disjoint paths. RIN provides higher reliability values than 3 disjoint IEGN [58], 2 disjoint MINs and scalable crossbar [56] and also provides four disjoint paths with same cost layout and also comparable reliability values with IEGN [58].

The reliability values of various MINs are evaluated and compared with proposed designs. MSMT and All - Terminal Reliability values of RIN are found to be higher than all 4 disjoint MINs, 3 disjoint MINs and 2 disjoint MINs. Reliable interconnection networks (RIN) provides higher MSMT reliability values than some recently proposed MINs [45–55] and provide higher reliability values for lower node reliability values (r = 0.50) compared to other MINs.

Table 5.12 Multistage interconnection networks broadcast reliability evaluation.

Network types	Broadcast Rel for Switch Reliability (r) =		
	0.99	**0.90**	**0.50**
Banyan [31], Baseline, Reverse Baseline, Reverse Omega [32], Delta [33], Cube [34], Omega [35], Shuffle Exchange [36]	0.9320	0.4783	0.0078
Butterfly [37]	0.8600	0.2059	0.0061
Dot network [38]	0.9135	0.3874	0.0020
Gamma [41]	0.8948	0.2965	1.2×10^{-4}
Monogamma GIN [44]	0.8944	0.2882	9.15×10^{-5}
Theta [39]	0.8594	0.1921	7.63×10^{-6}
Scalable crossbar [66]	0.9606	0.6561	0.0625
Augmented SEN [36]	0.9118	0.3246	3×10^{-5}
PHI [40]	0.9105	0.2984	1.94×10^{-5}
Extra Stage GIN [42]	0.9131	0.3662	1.8×10^{-4}
Fully chained GIN [47]	0.9128	0.3525	1×10^{-4}
Partial chained GIN [47]	0.9130	0.3615	1.94×10^{-4}

(Continued)

Table 5.12 Multistage interconnection networks broadcast reliability evaluation. (*Continued*)

Network types	Broadcast Rel for Switch Reliability (r) =		
	0.99	**0.90**	**0.50**
No Backtracking GIN [48], Minimal links traversed GIN [50], Combining switches MIN [51]	0.9126	0.3527	2.1×10^{-4}
Balanced Gamma [52]	0.8950	0.3014	1.54×10^{-4}
Benes [57]	0.9503	0.5523	0.0061
Extra-stage Benes [67]	0.9503	0.5498	0.0045
Extra stage SEN (SEN+) [71], Extra Stage Cube [58]	0.9503	0.5548	0.0083
Parallel Benes [67]	0.9982	0.8353	0.0086
SEN+2 and Omega +2 [71]	0.9507	0.5670	0.0074
Kappa network [56]	0.9129	0.3647	3.47×10^{-4}
Replicated Benes [67]	0.9986	0.8668	0.0170
Extra group network [68]	0.9133	0.3468	3.71×10^{-5}
3-disjoint paths GIN [49]	0.9134	0.3765	3.8×10^{-4}
Enhanced IADM [49]	0.9135	0.3864	0.0013

(*Continued*)

Table 5.12 Multistage interconnection networks broadcast reliability evaluation. (*Continued*)

Network types	**Broadcast Rel for Switch Reliability (r) =**		
	0.99	**0.90**	**0.50**
Pars Network [69]	0.9133	0.3695	5.95×10^{-5}
Improved extra group [68]	0.9137	0.3729	1.13×10^{-4}
Extended Generalized SEN [63]	0.9139	0.3856	1.73×10^{-4}
Incomplete GIN [55]	0.8948	0.2946	4×10^{-5}
Generalized SEN [60]	0.9042	0.3452	7.32×10^{-4}
Augmented Baseline [59]	0.9122	0.3382	6.1×10^{-5}
Tandem Omega [62]	0.9123	0.3333	1.7×10^{-5}
Two-dimensional ring Banyan [61]	0.9131	0.3717	4.3×10^{-4}
Clos Network [64]	0.9509	0.5904	0.0293
Modified Baseline [65]	0.7936	0.0886	1.1×10^{-7}
Proposed 4-Disjoint Multistage Interconnection Network (4DMIN)	**0.9509**	**0.5895**	**0.0189**
Reliable Interconnection Network (RIN)	**0.9995**	**0.9419**	**0.0758**

Table 5.13 Multistage interconnection networks all terminal reliability evaluation.

Network types	All Terminal Rel for Switch Reliability (r)		
	r = 0.99	**r = 0.90**	**r = 0.50**
Banyan [31], Baseline, Reverse Baseline, Reverse Omega [32], Delta [33], Cube [34], Omega [35], Shuffle Exchange [36]	0.8864	0.2824	2.45×10^{-4}
Butterfly [37]	0.7249	0.0344	1×10^{-6}
Dot network [38]	0.8775	0.2542	1.22×10^{-4}
Gamma [41]	0.7249	0.0344	1×10^{-6}
Monogamma GIN [44]	0.7845	0.0699	1×10^{-7}
Theta [39]	0.7528	0.0422	7.37×10^{-11}
Scalable crossbar [66]	0.8515	0.1853	1.53×10^{-5}
Augmented SEN [36]	0.8490	0.1406	5.2×10^{-8}
PHI [40]	0.8464	0.1219	1.9×10^{-8}
Extra Stage GIN [42]	0.8487	0.1364	3×10^{-8}
Fully chained GIN [47]	0.8511	0. 1766	2.5×10^{-6}
Partial chained GIN [47]	0.8511	0.1778	3.5×10^{-6}

(Continued)

Table 5.13 Multistage interconnection networks all terminal reliability evaluation. (*Continued*)

Network types	All Terminal Rel for Switch Reliability (r)		
	r = 0.99	**r = 0.90**	**r = 0.50**
No Backtracking GIN [48], Minimal links traversed GIN [50], Combining switches MIN [51]	0.8839	0.2236	2.7×10^{-6}
Balanced Gamma [52]	0.8491	0.1467	2×10^{-7}
Benes [57]	0.8504	0.1654	9×10^{-6}
Extra-stage Benes [67]	0.3866	0.9217	4×10^{-4}
Extra stage SEN (SEN+) [71], Extra Stage Cube [58]	0.9216	0.3844	2.89×10^{-4}
Parallel Benes [67]	0.9216	0.3887	5.4×10^{-4}
SEN+2 and Omega +2 [71]	0.9216	0.3887	1×10^{-4}
Kappa network [56]	0.9223	0.4067	5.56×10^{-4}
Replicated Benes [67]	0.8504	0.1653	9×10^{-6}
Extra group network [68]	0.8510	0.1411	1.24×10^{-8}
3-disjoint paths GIN [49]	0.8862	0.2623	1.1×10^{-5}
Enhanced IADM [49]	0.8514	0.1823	3.01×10^{-4}

(*Continued*)

Table 5.13 Multistage interconnection networks all terminal reliability evaluation. (*Continued*)

Network types	All Terminal Rel for Switch Reliability (r)		
	r = 0.99	**r = 0.90**	**r = 0.50**
Pars Network [69]	0.8512	0.1733	1.6×10^{-7}
Improved extra group [68]	0.8514	0.1738	1.75×10^{-7}
Extended Generalized SEN [63]	0.8519	0.1801	1.9×10^{-7}
Incomplete GIN [55]	0.8139	0.0796	3×10^{-10}
Generalized SEN [60]	0.8179	0.1216	1×10^{-6}
Augmented Baseline [59]	0.8502	0.1613	3.5×10^{-7}
Tandem Omega [62]	0.9224	0.4081	2.93×10^{-4}
Two-dimensional ring Banyan [61]	0.8507	0.1694	7.95×10^{-4}
Clos Network [64]	0.9228	0.4304	3.66×10^{-3}
Modified Baseline [65]	0.6173	0.00636	1×10^{-9}
4-Disjoint Multistage Interconnection Network (4DMIN)	**0.9227**	**0.4283**	**0.0012**
Reliable Interconnection Network (RIN)	**0.9991**	**0.9082**	**0.0160**

Table 5.13 Multi Source (S) Multi Terminal (T) Reliability Evaluation. (Node Reliability = 0.95).

Network types	1S/2D	2S/3D	3S/4D	4S/5D	5S/6D	6S/7D	7S/8D
Banyan [31], Baseline, Reverse Baseline, Reverse Omega [32], Delta [33], Cube [34], Omega [35], Shuffle Exchange [36]	0.8574	0.8145	0.7351	0.6302	0.5987	0.5688	0.5403
Butterfly [37]	0.7350	0.4632	0.3405	0.3073	0.2503	0.2259	0.2039
Dot network [38]	0.7737	0.7350	0.6302	0.5987	0. 5688	0.5403	0.5133
Gamma [41]	0.7682	0.6287	0.5121	0.4181	0.3405	0.2774	0.2259
Monogamma GIN [44]	0.8049	0.6527	0.5294	0.4538	0.4095	0.3770	0.2971
Theta [39]	0.6571	0.4702	0.4244	0.3804	0.3433	0.2533	0.2286
Scalable crossbar [66]	0.9025	0.7351	0.4401	0.4401	0.4401	0.4401	0.4401
Augmented SEN [36]	0.8403	0.7473	0.6745	0.6022	0.5379	0.4782	0.4316
PHI [40]	0.8209	0.6939	0.6160	0.5559	0.5017	0.4338	0.3915
Extra Stage GIN [42]	0.8543	0.7665	0.6860	0.6128	0.5463	0.4861	0.4319
Fully chained GIN [47]	0.8471	0.7539	0.6824	0.6114	0.5548	0.5022	0.4557
Partial chained GIN [47]	0.8501	0.7672	0.6871	0.5891	0.5570	0.5052	0.4563

(Continued)

Table 5.14 Multi Source (S) Multi Terminal (T) Reliability Evaluation (Node Reliability = 0.95). (*Continued*)

Network types	1S/2D	2S/3D	3S/4D	4S/5D	5S/6D	6S/7D	7S/8D
No Backtracking GIN [48], Combining switches MIN [51]	0.8960	0.7989	0.7505	0.6727	0.6306	0.5691	0.5339
Minimal links traversed GIN [50]	0.8471	0.7539	0.6774	0.6031	0.5372	0.4849	0.4348
Balanced Gamma [52]	0.7699	0.6271	0.5095	0.4587	0.4139	0.3736	0.3043
Benes [57]	0.8935	0.8488	0.8064	0.7606	0.7146	0.6789	0.6449
Extra-stage Benes [67]	0.8931	0.8484	0.8002	0.7516	0.7140	0.6783	0.6444
Extra stage SEN (SEN+) [71], Extra Stage Cube [58]	0.8939	0.8492	0.8067	0.7611	0.7152	0.6794	0.6454
Parallel Benes [67]	0.8341	0.7398	0.6561	0.5855	0.5010	0.4789	0.4322
SEN+2 and Omega +2 [71]	0.8971	0.8523	0.8071	0.7642	0.7260	0.6897	0.6552
Kappa network [56]	0.8509	0.7679	0.6863	0.6150	0.5549	0.4997	0.4499
Extra group network [68]	0.8490	0.7535	0.6709	0.6007	0.5421	0.4893	0.4415
3-disjoint paths GIN [49]	0.9018	0.8138	0.7719	0.6600	0.6254	0.5934	0.5629
Enhanced IADM [49]	0.8572	0.7733	0.6978	0.6295	0.5681	0.5125	0.4625

(*Continued*)

Table 5.14 Multi Source (S) Multi Terminal (T) Reliability Evaluation (Node Reliability = 0.95). (*Continued*)

Network types	**1S/2D**	**2S/3D**	**3S/4D**	**4S/5D**	**5S/6D**	**6S/7D**	**7S/8D**
Pars Network [69]	0.8548	0.7683	0.6926	0.6252	0.5642	0.5092	0.4595
Improved extra group [68]	0.8549	0.7696	0.6932	0.6255	0.5650	0.5100	0.4600
Extended Generalized SEN [63]	0.8573	0.7736	0.7106	0.6366	0.5770	0.5211	0.4638
Incomplete GIN [55]	0.8545	0.7673	0.6873	0.6145	0.5479	0.4875	0.4329
Generalized SEN [60]	0.8145	0.7737	0.6983	0.5987	0.4876	0.4632	0.4401
Augmented Baseline [59]	0.8403	0.7473	0.6745	0.6022	0.5379	0.4782	0.4316
Tandem Omega [62]	0.8973	0.8501	0.8052	0.7650	0.7267	0.6904	0.6559
Two-dimensional ring Banyan [61]	0.8570	0.7694	0.6983	0.6217	0.5584	0.5025	0.4536
Clos Network [64]	0.9024	0.8573	0.8145	0.7738	0.7351	0.6983	0.6634
Modified Baseline [65]	0.8145	0.4632	0.3774	0.3074	0.2774	0.2259	0.2039
4-Disjoint Multistage Interconnection Network (4DMIN)	**0.9025**	**0.8573**	**0.8144**	**0.7736**	**0.7349**	**0.6982**	**0.6632**
Reliable Interconnection Network (RIN)	**0.9948**	**0.9893**	**0.9868**	**0.9814**	**0.9800**	**0.9792**	**0.9786**

Table 5.15 Multi Source (S) Multi Terminal (T) Reliability Evaluation (Node Reliability = 0.50).

Network types	1S/2D	2S/3D	3S/4D	4S/5D	5S/6D	6S/7D	7S/8D
Banyan [31], Baseline, Reverse Baseline, Reverse Omega [32], Delta [33], Cube [34], Omega [35], Shuffle Exchange [36]	0.125	0.0625	0.0156	0.0020	9.7×10^{-4}	4.8×10^{-4}	2.44×10^{-4}
Butterfly [37]	0.016	3×10^{-5}	4.7×10^{-7}	1.1×10^{-7}	7.4×10^{-9}	1.8×10^{-9}	4.6×10^{-10}
Dot network [38]	0.031	0.015	0.002	9.7×10^{-4}	4.8×10^{-4}	2.4×10^{-4}	1×10^{-4}
Gamma [41]	0.0156	0.0015	9.15×10^{-5}	7.63×10^{-6}	4.76×10^{-7}	2.98×10^{-8}	1.86×10^{-9}
Monogamma GIN [44]	0.023	0.001	6.3×10^{-5}	6.6×10^{-6}	1.6×10^{-6}	1.4×10^{-7}	1.1×10^{-8}
Theta [39]	0.002	4.8×10^{-6}	1.2×10^{-6}	1.82×10^{-7}	4.5×10^{-8}	7.7×10^{-10}	1.9×10^{-10}
Scalable crossbar [66]	0.25	0.0156	0.0039	1.52×10^{-5}	1.52×10^{-5}	1.52×10^{-5}	1.52×10^{-5}
Augmented SEN [36]	0.0192	0.0016	4×10^{-4}	3.7×10^{-5}	4.6×10^{-6}	4×10^{-7}	1×10^{-7}
PHI [40]	0.005	2.9×10^{-4}	2.6×10^{-5}	6.6×10^{-6}	1.6×10^{-6}	1.5×10^{-7}	3.8×10^{-8}
Extra Stage GIN [42]	0.0459	0.006	6.95×10^{-4}	$7.5x 10^{-5}$	7.99×10^{-6}	7.45×10^{-7}	8.66×10^{-8}
Fully chained GIN [47]	0.026	0.001	3.9×10^{-4}	4.9×10^{-5}	2.01×10^{-5}	7.1×10^{-6}	1.14×10^{-7}

(*Continued*)

Table 5.15 Multi Source (S) Multi Terminal (T) Reliability Evaluation (Node Reliability = 0.50). *(Continued)*

Network types	1S/2D	2S/3D	3S/4D	4S/5D	5S/6D	6S/7D	7S/8D
Partial chained GIN [47]	0.031	0.007	0.0011	1.56×10^{-4}	5.1×10^{-5}	1.4×10^{-5}	4.5×10^{-6}
NBGIN [48], Combining switches MIN [51]	0.125	0.01	0.0027	4.3×10^{-4}	8.15×10^{-5}	2.04×10^{-5}	5.4×10^{-6}
Minimal links traversed GIN [50]	0.041	0.004	7.4×10^{-4}	6.6×10^{-5}	7.5×10^{-6}	1.8×10^{-6}	3.4×10^{-7}
Balanced Gamma [52]	0.018	0.0011	5.14×10^{-5}	$9.6x 10^{-6}$	2.4×10^{-6}	6×10^{-7}	3.8×10^{-8}
Benes [57]	0.085	0.042	0.021	0.006	0.0016	7.8×10^{-4}	3.9×10^{-4}
Extra-stage Benes [67]	0.0653	0.0327	0.0091	0.0023	0.0012	5.8×10^{-4}	2.89×10^{-4}
SEN+ [71], Extra Stage Cube [58]	0.109	0.054	0. 027	0.008	0.002	0.001	5.3×10^{-4}
Parallel Benes [67]	0.0051	2.5×10^{-4}	1.3×10^{-5}	8.9×10^{-5}	6.2×10^{-6}	1.2×10^{-7}	3×10^{-8}
SEN+2 and Omega +2 [71]	0.0927	0.0463	0.0147	0.0044	0.0023	0.0011	0.0005
Kappa network [56]	0.053	0.013	0.012	1.5×10^{-4}	3.8×10^{-5}	8.4×10^{-6}	1.8×10^{-6}
Extra group network [68]	0.0120	2.1×10^{-4}	1.3×10^{-5}	1.6×10^{-6}	3.9×10^{-7}	9.9×10^{-8}	2.48×10^{-8}

(Continued)

Table 5.15 Multi Source (S) Multi Terminal (T) Reliability Evaluation (Node Reliability = 0.50). (*Continued*)

Network types	**1S/2D**	**2S/3D**	**3S/4D**	**4S/5D**	**5S/6D**	**6S/7D**	**7S/8D**
3-disjoint paths GIN [49]	0.1562	0.0344	0.0107	6.3×10^{-4}	1.7×10^{-4}	6.7×10^{-5}	2.7×10^{-5}
Enhanced IADM [49]	0.096	0.017	0.004	7.5×10^{-4}	1.57×10^{-4}	3.24×10^{-5}	6.83×10^{-6}
Pars Network [69]	0.0218	5.8×10^{-4}	8.93×10^{-5}	2.03×10^{-5}	5.6×10^{-6}	1.4×10^{-6}	3.5×10^{-7}
Improved extra group [68]	0.0232	0.0012	1.3×10^{-4}	2.1×10^{-5}	6.7×10^{-6}	1.3×10^{-6}	3×10^{-7}
Extended Generalized SEN [63]	0.0479	0.0031	0.0010	8.1×10^{-4}	2×10^{-4}	3.3×10^{-5}	1.1×10^{-5}
Incomplete GIN [55]	0.05	0.009	0.001	1.2×10^{-4}	1.18×10^{-5}	1.13×10^{-6}	1.14×10^{-7}
Generalized SEN [60]	0.0625	0.0313	0.0078	9.7×10^{-4}	6.1×10^{-5}	3.05×10^{-5}	1.5×10^{-5}
Augmented Baseline [59]	0.0300	0.0032	7.95×10^{-4}	7.62×10^{-5}	1.9×10^{-5}	2.8×10^{-6}	7×10^{-7}
Tandem Omega [62]	0.0482	0.0151	0.0047	0.0023	0.0011	5.9×10^{-4}	2.9×10^{-4}
Clos Network [64]	0.234	0.1171	0.0586	0.0293	0.015	0.007	0.003
Modified Baseline [65]	0.0625	3×10^{-5}	1.9×10^{-6}	1.1×10^{-7}	2.9×10^{-8}	1.8×10^{-9}	4.6×10^{-10}
4DMIN	**0.1955**	**0.0818**	**0.0370**	**0.0152**	**0.0058**	**0.0029**	**0.0012**
RIN	**0.3585**	**0.1134**	**0.0703**	**0.0266**	**0.0215**	**0.0188**	**0.0159**

It should be noted that in the field of interconnection networks, there are a large number of network topologies, each with their own specific features and structures. Therefore, various designs of MINs are designed to be fault-tolerant without increasing its hardware complexity or even with its cost reduced, to achieve a new network with high reliability.

The choice of MINs that may vary with different engineering applications and choices of design may vary from different aspects. Various designs are discussed in literatures. However, all these designs are valuable in the field of interconnection networks. These literatures have been focused to improve the fault tolerance and reliability of MINs. From our observation our proposed designs (4DMINs and RINs) surpasses other MINs and hence it makes a perfect choice for multiprocessor interconnection networks.

5.4 Summary

This chapter has attempted to evaluate and provide comprehensive reliability paradigms of various recent and industrial used MINs topologies available from literatures. This chapter presents a conceptually simple and computationally efficient approach for reliability evaluation of MINs with multiple sources and multiple destinations nodes. The versatility of the approach is that it can be used for evaluating all the network reliability parameters (2-TR, broadcast and ATR). Several benchmark networks available from literature have been examined and reliability values are evaluated. These evaluations are useful for designing high reliable MIN network layouts.

6

Dynamic Tolerant and Reliable Four Disjoint MIN Layouts

Interconnection networks are primarily utilized in tightly coupled multiprocessor systems and large-scale parallel/distributed systems for the purpose of broadband switching and inter-processor communications. These systems rely on reliable and fault tolerant multistage interconnection networks (MINs) to meet high performance requirements of large amounts of data transmission from a large set of source nodes to a large set of destination nodes. Multistage Interconnection Networks (MINs) are widely used in parallel and distributed systems, multiprocessor systems, communication networks, optical networks, modern embedded systems.

A network designer has to choose a suitable MIN architecture such that network is capable of providing high performance, high throughput, high availability and lower latency to route multiple communication tasks concurrently. To design such a network, a designer has to meet various performance requirements of the network like speed, latency, compactness, throughput, reliability etc.

Multistage switching fabrics are required to develop a fault tolerant, cost effective and low transmission delay multistage network. The fault tolerance and reliability of MIN architectures depend upon available disjoint and multiple paths. High reliability of MINs can be achieved using fault tolerance techniques. The fault-tolerance is generally achieved by disjoint paths available through multiple connectivity options.

Most of the discussion in MIN design is centered towards fault tolerance capability of designs. The basic idea for the fault-tolerance of the MIN is to create redundant (alternative) paths between all source-destination (S-D) node pairs. These redundant paths can be

Neeraj Kumar Goyal and S. Rajkumar. Interconnection Network Reliability Evaluation: Multistage Layouts, (157–202) © 2020 Scrivener Publishing LLC

used when any switch gets failed/busy. Information on redundant and disjoint paths provides better knowledge about network capability to withstand switch failures. However, networks with same fault tolerance capability may perform differently from reliability point of view, due to existence of redundant paths, which may not be completely disjoint.

A wide number of single disjoint path multistage interconnection networks are discussed in literature. The Banyan network [31], Baseline [32], Delta networks [33], Binary n-cube network [34], Omega network [35], Shuffle-exchange network [36], Butterfly [37] and Scalable crossbar network (SCN) [66] are some popular single disjoint path MIN architectures. These basic networks provide only one path between any source-destination node pairs using 2×2 switches as basic elements.

Gamma interconnection network (GIN) [41] is another class of inherent fault-tolerant MINs. From Table 6.3 we have noted that GIN provides only one path when source and destination node sequence number is same and provides multiple paths when the node pairs are different. Therefore, it provides different level of fault tolerance to different node pairs and fails to guarantee fault tolerance.

A wide number of two disjoint path multistage interconnection networks are available in literature. Benes Network [57], ASEN [36], Augmented Baseline Network [59], PHI network [40], PM22I interconnection network [75], extra stage gamma interconnection network [42], B-network GIN [43], partially chained and fully chained gamma interconnection network (PCGIN and FCGIN [47]), cyclic gamma interconnection network (CGIN) [46], No-backtracking Gamma Interconnection Network (NBGIN) [48] can provide at least 2-disjoint paths with almost same hardware cost as GIN.

Parallel Benes [67] and Extra-stage Benes [67] are some recent two disjoint MINs. Parallel Benes network (PBN) is constructed using multiple Benes networks in parallel. Extra-stage Benes network (EBN) is designed by increasing an extra stage to the original Benes network such that it provides additional paths between each source–destination pairs.

All these networks have been proposed to improve the fault tolerant capability of MIN with sole objective of increasing the redundant and disjoint paths, thereby increasing its reliability and fault-tolerance.

Some three disjoint path networks are: Three-Disjoint Gamma Interconnection Network (3DGIN) [49], Enhanced IADM [49] and Pars network [69]. This family of MINs has multiple redundant paths for source-destination node pairs. However, these possess minimum three disjoint paths for any source–destination node pair of the MIN.

4-disjoint path MINs like Non-blocking extended generalized shuffle (EGS) network [63] and Clos network [64] provide multiple redundant paths for source-destination node pairs. They offer better performance in terms of reliability and multicasting as these possess minimum four disjoint paths for any source-destination node pair of the MIN. EGS [63] is a class of strictly non-blocking multistage interconnection networks providing four disjoint paths but lacks re-routability at the intermediate stages. This network includes same size external splitters and combiners used for connecting inputs and outputs. But the inclusion of splitters, combiners and more switching elements increase hardware complexity and space requirements.

Salient features in designing a MIN [19, 79, 80] will require

a) Fault tolerant capability:
Min should provide multiple redundant and disjoint paths. Designs should have dynamic rerouting behavior capability to use the disjoint paths effectively. Dynamic rerouting provides beneficial re-routability in case of switch failures.

b) MINs should tolerate multiple switch components failures/busy states. This factor is a key to MIN fault tolerance capability because if one processor fails/busy, remaining processors can transfer communication without any interruption.

c) One of the most important aspects of performance that should be considered in design of MIN systems is reliability. Therefore, it is needed that the reliability of the MINs should be higher to determine the behavior of these networks.

d) Finally, the cost is a fundamental issue during MIN improvement. The cost is a very important parameter such that any design is not reasonable in practice if they possess very high cost layout.

Two new designs of 4-disjoint paths multistage interconnection networks called 4-disjoint multistage interconnection network and reliable interconnection network (4DMIN and RIN), are presented in this chapter. These layouts are multipath multi-stage interconnection networks (MIN) providing four disjoint paths for all the source-destination node pairs. These layouts show dynamic rerouting capability which further improves network fault-tolerant capability. Proposed designs are highly reliable MINs. Terminal pair reliabilities of proposed designs are evaluated and compared with other disjoint MINs. It is found that these designs successfully satisfy requirement of a high reliable and low-cost MIN layout. The designs have been evaluated and compared with existing MINs based on the cost.

6.1 Topological Design Considerations

4DMIN is a multipath MIN providing four disjoint paths. 4DMIN has full access property, i.e. it provides connection from any source nodes to any destination nodes. 4DMIN with the redundant paths is desirable for multiprocessor systems, as it reroute packets to other available alternative paths in case of faulty/busy path achieving enhanced reliability and performance.

The number of disjoint paths for each source-destination node pairs depends on the type of interconnection pattern and design topology. For same topology (same number/type of switches used), changing the interconnection patterns between the stages results in different design. Disjoint paths are due to first and last stage interconnection patterns while multiple paths are due to intermediate interconnection patterns between the stages.

4-disjoint paths multistage interconnection network named as 4DMIN design have *n+1* number of stages from Stage 0 to Stage n. Each stage involves N number of sources and N number of destinations where N is 2^n. The switch sizes and the number of switches in each stage of the network for proposed 4DMIN are given in the Table 6.4.

4DMIN is a highly fault tolerant network due to its multipath nature and non-blocking property. The proposed 4DMIN requires lesser hardware cost compared to other MINs of similar performance. Therefore, 4DMIN provide 4-disjoint paths and higher

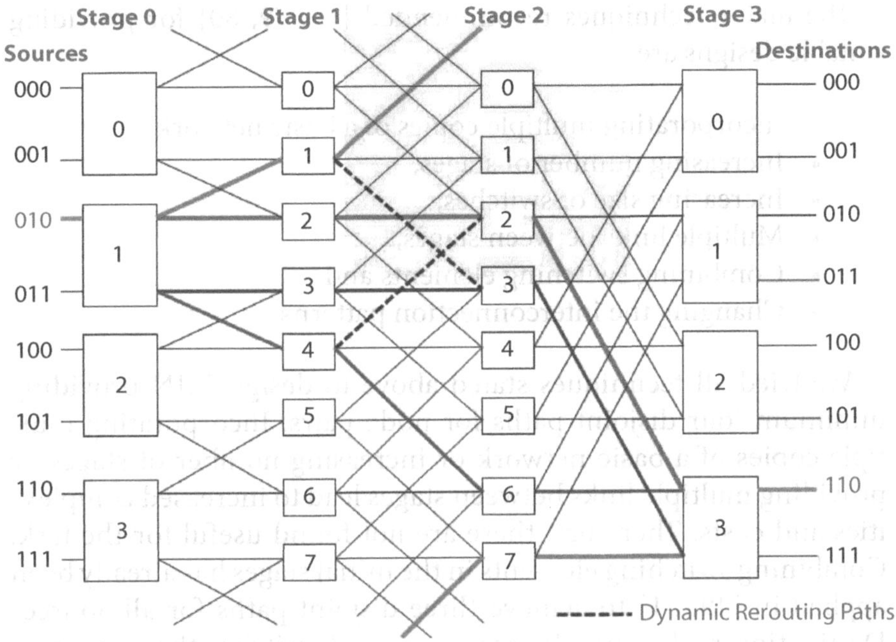

Figure 6.1 4DMIN showing four disjoints paths and six redundant paths for S-D node pair (2, 6; δ = 4).

reliability at lower hardware cost. The 4DMINwith marking of four disjoint paths is shown in Figure 6.1.

6.1.1 Topology

Proposed four stage (Stage 0-3) design of 4DMIN design involve 8 sources and 8 destinations ($N = 8 = 2^n = 2^3$). Four switching elements are used in the first stage (Stage 0) and last stage (Stage 3), which are 2×4 and 4×2 size crossbars, respectively. Number of disjoint paths for any source to any destination depends largely on the size of the initial and final stage switches. Therefore, it is required to use four outgoing link switches at Stage 0 and four incoming link switches at Stage 3.

The major considerations in designing four disjoint paths network are managing:

- Hardware complexity
- Design complexity

The major techniques recommended [19, 79, 80] for providing suitable designs are

- Incorporating multiple copies of a basic network,
- Increasing number of stages,
- Increasing size of switches,
- Multiple links between stages,
- Combining switching elements and
- Changing the interconnection patterns

We tried all techniques stated above to design MIN providing minimum four disjoint paths for node pairs. Incorporating multiple copies of a basic network or increasing number of stages or providing multiple links between stages lead to increased complexities and costs. Therefore, these are not found useful for the task. Combining switching elements in the initial stages has already been applied in [49, 51] to achieve three disjoint paths for all Source-Destination node pairs. Increasing size of switches through combination of switches provides more disjoint paths without much increase in design cost and complexities. Therefore, it has been used in proposed design. After carefully analyzing these factors, proposed 4DMIN are designed to provide higher fault tolerance and good performance. Proposed 4DMIN utilize same switch sizes in various stages.

6.1.2 Switch Selection for Proposed 4DMIN

In 4DMIN, Stage 3 (output stage) is mirror of Stage 0 (input stage) while intermediate stages, Stage 1 and Stage 2 are mirror images of each other. Therefore, Stage 0 is discussed along with Stage 3 and Stage 1 is discussed along with Stage 2.

Stage 0 & 3:
Initially, the problem formulation started with the objective of designing 4-disjoint path MIN. Therefore, switch size of Stage 0 has to be $Z \times 4$ as four output links are necessary to provide four disjoint paths. Value of Z has to be optimum, which can be any value out of 1, 2, 4 and 8. A lower value of Z ensures lower dependence and

simpler switch design but requires higher network chip size, power consumption and network cost, as more switches are required.

On the other hand, high value of Z increases probability of cross talk, interference, collision etc.; increases complexity of switch, increases number of connectivity losses as one input switch failure will cause Z number of disconnected input nodes but reduces network cost as number of switches required becomes lesser.

Choosing Z=2 makes a balance of the two. It reduces required number of switches (four switches are required in place of eight in case Z=1) and associated costs. It requires lesser complex switches reducing probability of cross talk, collisions etc. It allows network to provide good performance. Stage 3 is mirror image of Stage 0 and it uses four 4×2 switches. Following the same logic as input stage (Stage 0), the final Stage has to provide 4-disjoint paths and hence we have used four 4×2 switches, i.e. two nodes output switches are used.

Stage 1 & 2:
From Stage 0, sixteen links are emerging out. In order to evenly by-pass those links to Stage 1. We have use to use 2×Z size eight crossbar switches. Other option is to use 4×Z four crossbar switches. In proposed design 2×Z switches are used as 4×Z switch failures will lead to reduced fault tolerant capability.

Number of multiple paths depends largely on intermediate switch sizes. Higher number of multiple paths will be generated when Z is high. A higher Z provides higher redundancy, capacity and more alternate paths ensuring higher performance and reliability on switch/link failures.

However, it requires complex and costly switches and can increase chances of collision or crosstalk. Lower value of Z ensures simpler switch design, good link connectivity but provides lesser redundant paths resulting in lower available capacity between nodes, lower performance and lower reliability. On trying various values of Z, Z=3 was found to be making a balance between reliability and cost. It is observed that further increase in Z to 4 and above, improves reliability very little but requires higher cost. Therefore, eight switches of size 2×3 are used in Stage 1. Following similar logic as Stage 1 switch size selection, Stage 2 is mirror image of Stage 1 using eight 3×2 crossbars.

6.2 Proposed 4-Disjoint Multistage Interconnection Network (4DMIN) Layout

4-disjoint paths multistage interconnection network is a multipath IN providing four disjoint paths. It has full access property, i.e. it provides connection from any of its source nodes to any of its destination nodes.

4DMIN has full access and dynamic rerouting property under switch failure/busy states. When a message packet encounters any switch/link faults, link can change their routing path as per algorithm and thus handling collision problems effectively.

6.2.1 Switching Pattern

According to graph theory [81–83], a network system can be modeled as a graph G = {N, L} where N represents the set of nodes and L represents the set of links. In case of MINs, switches act as nodes connected by links. The proposed four stages (Stage 0-3) design of 8×8 size 4DMIN is shown in Figure 6.1.

Size of the MIN is represented by N(N=8)
Stages are represented as ***i***(i=0,1, 2 and 3)
Switches in every Stages are represented as ***j*** (j=0, 1, 2, 3, 4, 5, 6 and 7)

At Stage 0
Number of switches used in this stage is 4.

Stage *0* is constructed by using 2×4 size crossbars. So, every switch contains four output links. These four-output links connect switches at Stage *0* and Stage *1* in the following fashion:

- First link connects ((2j-1) mod N)th switch at Stage *1*
- Next link connects ((2j) mod N)th switch at Stage *1*
- Next link connects ((2j+1) mod N)th switch at Stage *1* and
- Fourth link connects ((2j+2) mod N)th switch at Stage *1*

At Stage 1:
Number of switches used in this stage is 8. Stage 1 is designed using 2×3 size crossbars. So, every switch contains three output links connecting to switches at Stage 1 and Stage 2 based on the plus-minus

2^i functions. These three-output links connect switches at Stage *1* and Stage *2* in the following order:

- First link connects $[(j-2^i) \bmod N]^{th}$ switch at Stage *2*
- Next link connects j^{th} switch at Stage *2* and
- Final link connects $[(j+2^i) \bmod N]^{th}$ switch at Stage *2*.

At Stage 2:
Eight switches are used in this stage. Stage 2 is designed by using 3×2 size crossbars. Therefore, every switch contains only two output links connecting switches at Stage 2 and Stage 3.

6.2.2 Redundant and Disjoint Paths

Proposed layout provides four disjoint paths for all the source (S) and destination (D) node pairs. This design also provides more redundant paths for every S-D node pairs. But number of redundant paths for every S-D node pairs varies depending on their tag value (**δ**) and shown in Table 6.1.

Tag value can be calculated using difference modulo N function between source and destination numbers. The relative position of nodes can be represented through tag value (**δ**), which is the difference between source and destination numbers **δ=|D-S|**.

Now consider the Figure 6.1, for the flow of data from source node **2 (010)** to destination node **6 (110)**. The difference between S and D gives the tag value **δ=|6-2|** is **4**. A wide number of MINs provides different number of redundant paths based on their tag value.

Gamma interconnection network (GIN) [41], extra stage gamma interconnection network (ESGIN), [42], B-network GIN [43], partially chained and fully chained gamma interconnection network

Table 6.1 Total number of paths for every tag for 8×8 4DMIN-1.

	Tag value (δ)							
Redundant Paths	**0**	**1**	**2**	**3**	**4**	**5**	**6**	**7**
4DMIN	6	6	6	6	6	6	6	6
Disjoint Paths	4	4	4	4	4	4	4	4

(PCGIN and FCGIN [47]), cyclic gamma interconnection network (CGIN) [46], balanced gamma interconnection network (BGIN) [52], three-disjoint gamma interconnection network (3DGIN) [49], three-disjoint cyclic gamma interconnection network (3D-CGIN) [53] provide variable alternative paths for different tag value and shown in Table 6.2 and Table 6.3.

6.2.3 Routing and Dynamic Rerouting

MINs switch can route packets using distance routing and destination routing method. In distance routing method, routing tags are generated from difference modulo N function between source node and destination node pairs. Here routing tags have to be calculated every time for every message transmission. To overcome this problem destination-tag routing is commonly used. In destination-tag routing, message routing is carried out through destination address. Here switch settings are determined solely by the message destination address $(z_0\, z_1\, z_2 \ldots z_{n-1})$.

4DMIN uses destination-based routing method to route packets. Outgoing links and the destination bits representation are shown in the Figure 6.2.

Table 6.2 Total number of alternative paths for every tag in 8×8 MINs.

	Tag values (δ)							
Network	**0**	**1**	**2**	**3**	**4**	**5**	**6**	**7**
GIN	1	4	3	5	2	5	3	4
ESGIN	9	8	12	10	12	10	12	8
CGIN	5	4	4	2	2	2	4	4
PCGIN	18	18	18	18	18	18	18	18
FCGIN	64	64	64	64	64	64	64	64
3D-CGIN	7	6	8	6	7	6	8	6
3DGIN	6	6	6	5	4	4	5	5
4DMIN	**6**	**6**	**6**	**6**	**6**	**6**	**6**	**6**

Table 6.3 Number of disjoint paths for every tag in 8×8 MINs.

Network	Tag values (δ)							
	0	**1**	**2**	**3**	**4**	**5**	**6**	**7**
GIN	0	2	0	2	0	2	0	2
ES-GIN	3	3	3	3	3	3	3	3
CGIN	3	2	3	2	2	2	3	2
PCGIN	2	2	2	2	2	2	2	2
FCGIN	2	2	2	2	2	2	2	2
3D-CGIN	3	3	3	3	3	3	3	3
3DGIN	3	3	3	3	3	3	3	3
4DMIN	**4**	**4**	**4**	**4**	**4**	**4**	**4**	**4**

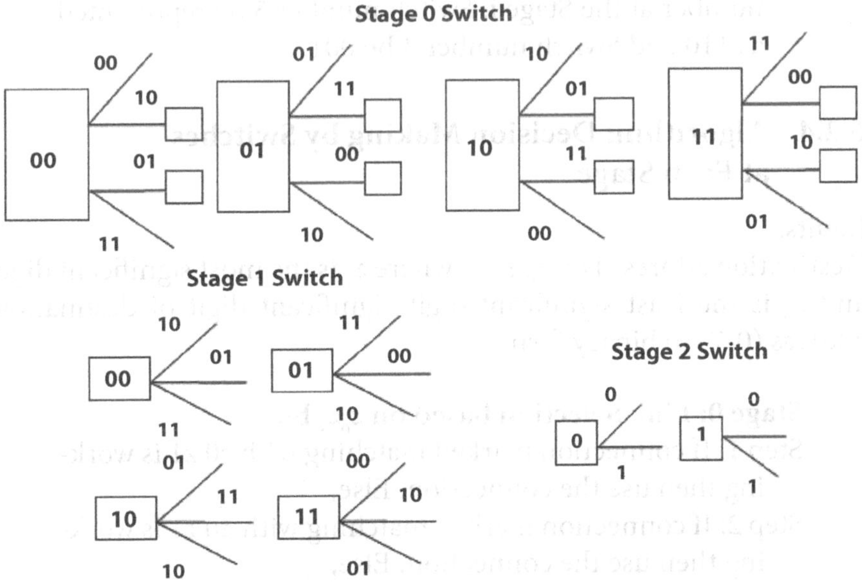

Figure 6.2 Outgoing Links and the Routing Bits from Stage 0 to Stage 2 based on Destination Bits.

Switch/link selection for 8×8 4DMINis based on:

- Stage 0 decision rely on $z_0\,z_1$ bits
- Stage 1 selection based on $z_1\,z_2$ bits
- Stage 2 selections through z_2 bits

Notations:

j = Number of Switches in a Stage i
k = Binary representation of switch number at the Stage i.
z = Destination address bits
N = Number of Source-Destination nodes (N = 8) Where n = $\log_2$N
i = Number of Switching Stages where i = (n+1) Where i = 0, 1, 2 and 3

- $j = j_0\,j_1$........j_{n-1} be the switches at Stage 1-3. Thus, j_0 j_1........j_7 be the switches at Stages 1 and 2 and $j_0\,j_1\,j_2\,j_3$ switches at Stages 3.
- Destination bits are represented as $z = z_0\,z_1\,z_2 \ldots z_{n-1}$.
- $k_0\;k_1k_2 \ldots k_{n-1}$ be the binary representation of switch number at the Stage i. Switch number 3 be represented as 110 and Switch number 4 be 001.

6.2.4 Algorithm: Decision Making by Switches at Each Stage

Inputs:
Destination address D = $z_0\,z_1\,z_2$, where z_2 is the most significant digit and z_0 is the least significant digit significant digit of destination address (0-7) in binary form.

Stage 0: Link Selection based on z_0z_1 bits
Step 1: If connection marked matching with $z0\;z1$ is working then use the connection. Else,
Step 2: If connection marked matching with $\overline{z0}\;z1$ is working then use the connection. Else,
Step 3: If connection marked matching with $z0\;\overline{z1}$ is working then use the connection. Else,

Step 4: If connection marked matching with $\overline{z0z1}$ is working then use the connection. Else,

Step 5: Links are Busy/Failed. Connection cannot be made.

$$\left(k_0, z_0z_1\right) = \begin{cases} (2j-1) \bmod N & \text{if } k_0 = 0 \text{ and } z_0z_1 = 10 \text{ or if } k_0 = 1 \text{ and } z_0z_1 = 11 \\ (2j) \bmod N & \text{if } k_0 = 0 \text{ and } z_0z_1 = 01 \text{ or if } k_0 = 1 \text{ and } z_0z_1 = 00 \\ (2j+1) \bmod N & \text{if } k_0 = 0 \text{ and } z_0z_1 = 11 \text{ or if } k_0 = 1 \text{ and } z_0z_1 = 10 \\ (2j+2) \bmod N & \text{if } k_0 = 0 \text{ and } z_0z_1 = 00 \text{ or if } k_0 = 1 \text{ and } z_0z_1 = 01 \end{cases}$$

Stage 1: Link Selection based on z_1z_2 bits

Step 1: If connection marked matching with $z1z2$ is working then use the connection. Else,

Step 2: If connection matching with $z1\ \overline{z2}$ is available then use the connection.

$$\left(k_ik_{i+1}, z_iz_{i+1}\right) = \begin{cases} & \text{if } k_ik_{i+1} = 00 \text{ and } z_iz_{i+1} = 10, \text{ if } k_ik_{i+1} = 10 \text{ and } z_iz_{i+1} = 01 \\ (j-2^i) \bmod N & \text{if } k_ik_{i+1} = 01 \text{ and } z_iz_{i+1} = 11, \text{ if } k_ik_{i+1} = 11 \text{ and } z_iz_{i+1} = 00 \\ j & \text{if } k_ik_{i+1} = 00 \text{ and } z_iz_{i+1} = 01, \text{ if } k_ik_{i+1} = 10 \text{ and } z_iz_{i+1} = 11 \\ & \text{if } k_ik_{i+1} = 01 \text{ and } z_iz_{i+1} = 00, \text{ if } k_ik_{i+1} = 11 \text{ and } z_iz_{i+1} = 10 \\ (j+2^i) \bmod N & \text{if } k_ik_{i+1} = 00 \text{ and } z_iz_{i+1} = 11, \text{ if } k_ik_{i+1} = 10 \text{ and } z_iz_{i+1} = 00 \\ & \text{if } k_ik_{i+1} = 01 \text{ and } z_iz_{i+1} = 10, \text{ if } k_ik_{i+1} = 11 \text{ and } z_iz_{i+1} = 01 \end{cases}$$

$$\left(k_0k_1, z_1z_2\right) = \begin{cases} & \text{if } k_0k_1 = 00 \text{ and } z_1z_2 = 10, \text{ if } k_0k_1 = 10 \text{ and } z_1z_2 = 01 \\ (j-2^i) \bmod N & \text{if } k_0k_1 = 01 \text{ and } z_1z_2 = 11, \text{ if } k_0k_1 = 11 \text{ and } z_1z_2 = 00 \\ j & \text{if } k_0k_1 = 00 \text{ and } z_1z_2 = 01, \text{ if } k_0k_1 = 10 \text{ and } z_1z_2 = 11 \\ & \text{if } k_0k_1 = 01 \text{ and } z_1z_2 = 00, \text{ if } k_0k_1 = 11 \text{ and } z_1z_2 = 10 \\ (j+2^i) \bmod N & \text{if } k_0k_1 = 00 \text{ and } z_1z_2 = 11, \text{ if } k_0k_1 = 10 \text{ and } z_1z_2 = 00 \\ & \text{if } k_0k_1 = 01 \text{ and } z_1z_2 = 10, \text{ if } k_0k_1 = 11 \text{ and } z_1z_2 = 01 \end{cases}$$

Stage 2: Link Selection based on z_2 bits

Choose the switch/link by z_2.

$$\left(k_2, z_2\right) = \begin{cases} (j-2^i)\bmod N \text{ or } (j+2^i)\bmod N & \text{if } k_2 = 0 \text{ and } z_2 = 1, \text{ if } k_2 = 1 \text{ and } z_2 = 0 \\ j & \text{if } k_2 = 0 \text{ and } z_2 = 0, \text{ if } k_2 = 1 \text{ and } z_2 = 1 \end{cases}$$

6.2.5 Case Example

Consider the network size N=8 considering Source S=3 and Destination=6. So, the routing tag = ***z0 z1 z2*** = **011**. The routing of 4DMIN is shown in Figure 6.3.

Case 1: No switch failure at all the Stages:

Stage 0: *z*0*z*1 (01) so use the switch 4 at Stage 1.
Stage 1: *z*1*z*2 (11) so use the switch 2 at Stage 2.
Stage 2: choose the switch 3 at Stage 3 by *z*2 (1).

Case 2: Switch modes at Stage 2:
When switch 2 (**z1z2** (11)) busy/fails at Stage 2, the algorithm will enable switch 4 (**z0z1**(01)) at Stage 1 to choose switch 6 ($\mathbf{z1\overline{z2}}$(10)) as the dynamic rerouting path at Stage 2 and finally switch 3(**z2**(1)) at Stage 3 is chosen.

- Considering there is no busy/failures at Stage 1 but switch 6 busy/fails at Stage 2, then there is no use of switch 4 at Stage 1.
- In this condition, the algorithm automatically selects switch 1 ($\mathbf{\overline{z0}z1}$(11)) at Stage 1 as the next link. Next link bypasses the data *via* switch 3(**z1z2**(11)) in Stage 2.
- Again, if the switch 3 busy/fails at Stage 2 then the routing algorithm routes the data through dynamic rerouting path by switch 7 ($\mathbf{z1\overline{z2}}$(10)) at Stage 2.
- Even if the switch 7 busy/fails, the algorithm will not able to send data because the design can withstand only 3 switch failures at Stage 2.

Case 3: Switch Operational Modes at Stage 1:

- It is already known that the algorithm selects switch 4 (**z0z1**(01)) at Stage 1 by default for routing packets in this example. For instance, when switch 4 busy/fails at the very beginning Stage, then switch 1 (**z0z1**(11)) is chosen.
- Likewise, even if switch 1 busy/fails, algorithm will choose switch 2 (**z0z1**(00)) and finally switch 3 (**z0z1**(10)) is chosen on the busy/failure of switch 2.
- If the switch 3 also busy/fails at the first Stage, the algorithm may not able to find another path to the destination.

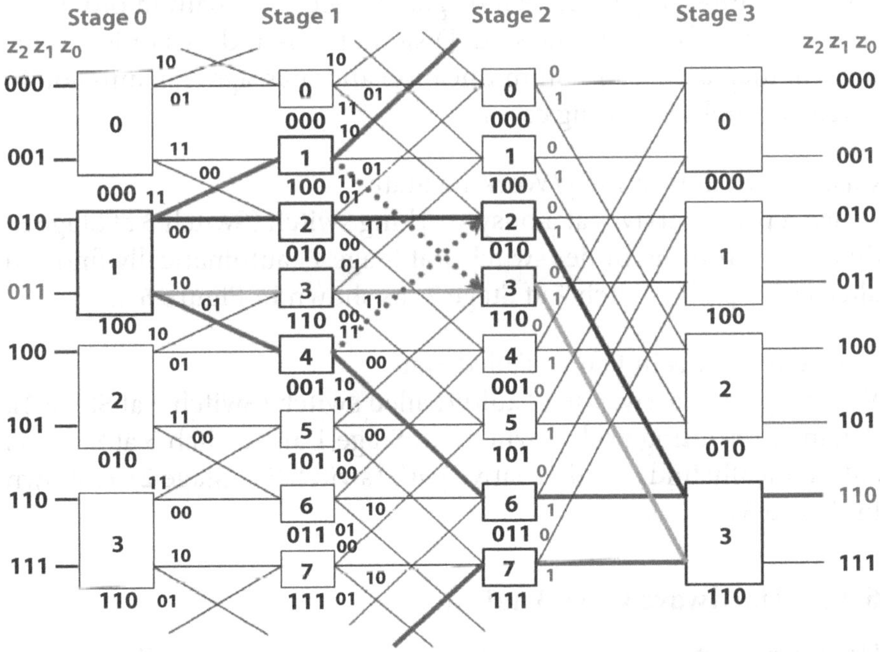

Figure 6.3 Dynamic rerouting concept in 4DMIN for S-D node pair (3, 6; δ = 3).

6.2.6 Disjoint and Dynamic Rerouting Approach in 4DMIN

It is an alternative approach to design a fault tolerant MIN under switch/link failures without backtracking. It corrects the routing tag in which the transmission changes the routing path to find an alternative path once a faulty or busy element is met. When a message packet encounters any switch/link faults, they change their routing path and effectively handle collision problems.

A switching element (SE) can have only two states (i) working or (ii) failed. Disjoint paths and dynamic rerouting behavior improve MIN fault tolerant capability. When a packet arrives at a switch, its header contains source and destination memory address, routing information, control information and sequence number (SN).

The sequence number is needed to re-order the packets of a message if they may get out of order in transit. This may occur, for example, if different packets follow different paths between the source and the destination. If packets can be guaranteed to remain in order, the sequence number is not needed. Disjoint paths and dynamic rerouting will increase faster communication and message transmission in MINs through preceding cases.

Case 1: Switch is in Busy/Working State
When a packet arrives at a busy/working switch (Switch 3 at Stage 2) dynamic rerouting node (switch 1 at Stage 1) automatically finds an alternative path (switch 7 at Stage 2) as shown in Figure 6.4.

Case 2: Switch is in Faulty/Failed State
When a packet arrives at a Faulty/Failed switch (Switch 3 at Stage 2), dynamic rerouting node (switch 1 at Stage 1 and switch 5 at Stage 1) automatically finds an alternative path (switch 7 at Stage 2) as shown in Figure 6.5.

6.2.7 Hardware Cost Analysis

Hardware cost analysis of various MIN architectures is discussed in Section 2.7 and provided in Table 2.2. The cost of a network [74] is calculated using design details such as number of components

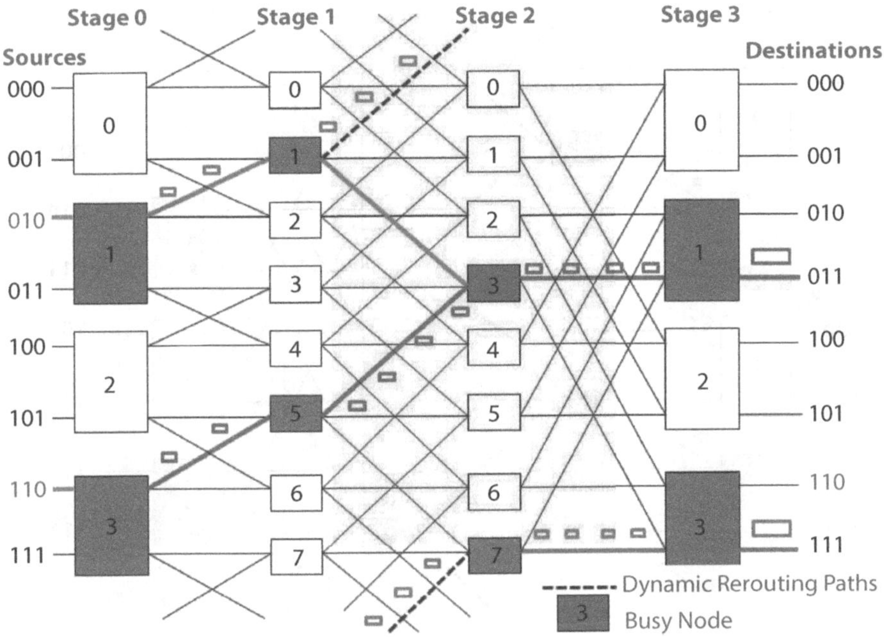

Figure 6.4 Dynamic rerouting behavior during Busy/Working switch.

(switches and links) used in the network, and number of connection points.

Hardware complexity of 4DMIN is lesser than other 2-disjoint and 3-disjoint path MINs. Hardware analysis for different tag MINs are calculated and shown in Table 6.4. A sample cost calculation is shown below for proposed 4DMIN:

Total no of 2×4 switching elements = 4 giving cost = I×O×S
= 2×4×4 = 32

Total no of 2×3 switching elements = 8 giving cost = I×O×S
= 2×3×8 = 48

Total no of 3×2 switching elements = 8 giving cost = I×O×S
= 3×2×8 = 48

Total no of 4×2 switching elements = 4 giving cost = I×O×S
= 4×2×4 = 32

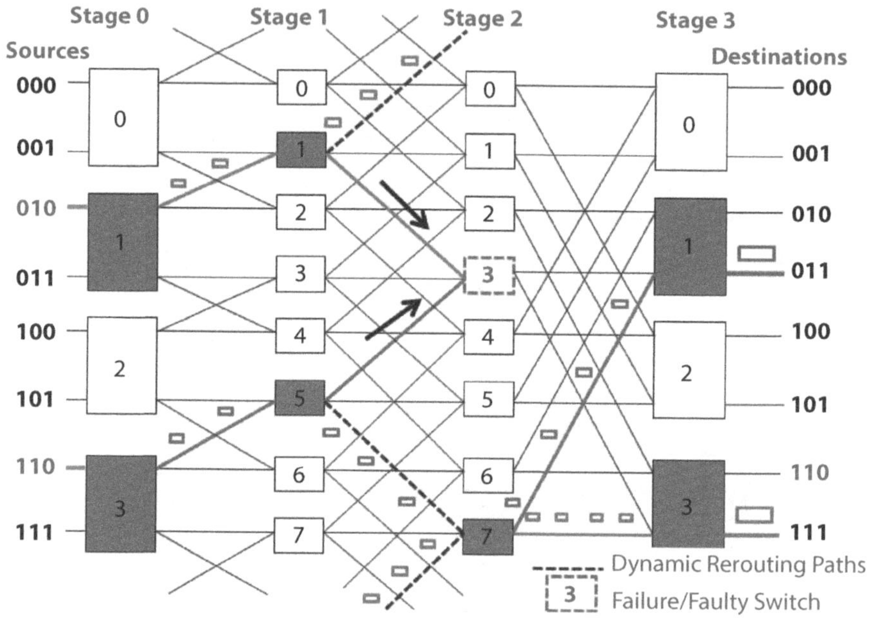

Figure 6.5 Dynamic rerouting behavior during Faulty/Failed switch.

Total cost of 4DMIN for N=8 is 32 + 48+48+32 = **160** units.

Total cost of switches is calculated for all the MINs as product of number of input pins (I), output pins (O) and switches (S).

6.3 Reliability Analysis and Comparison of MINs

Family of GINs and 4DMIN provides variable alternative paths for different tag value. In this section terminal reliability, which varies depending on the tag value of various MINs is analyzed. Following general assumptions are made for the terminal reliability evaluation in this work:

- All failures are statistically independent
- Links of the network are perfectly reliable
- The switch has only two states (i) working or (ii) failed.

Table 6.4 Size of crossbar switches of various MIN architectures for N=8.

Stage no	Crossbar switches [Input Pins(I) x Output Pins(O) x (Switches(S))]								
	GIN	**ESGIN**	**CGIN**	**PCGIN**	**FCGIN**	**3DCGIN**	**3DGIN**	**4DMIN**	**RIN**
0	1×3×(8)	1×3×(8)	1×3×(8)	2×4×(8)	2×3×(8)	2×3×(8)	2×4×(4)	2×4×(4)	2×4×(8)
1	3×3×(8)	3×3×(8)	3×3×(8)	3×3×(8)	3×3×(8)	3×3×(8)	2×3×(8)	2×3×(8)	4×3×(8)
2	3×3×(8)	3×3×(8)	3×3×(8)	3×2×(8)	3×3×(8)	3×3×(8)	3×3×(8)	3×2×(8)	3×2×(8)
3	3×1×(8)	3×3×(8)	3×1×(8)	2×1×(8)	2×1×(8)	3×1×(8)	3×1×(8)	4×2×(4)	2×2×(8)
4	–	3×1×(8)	–	–	–	–	–	–	
Cost=Σ(I×O×S)	192	264	192	200	208	216	176	160	240

- Partial functionality loss of switches is considered as failure of switch.

Terminal reliability is an important performance measure of MIN. It is calculated using simple, general and efficient path-based method discussed in Chapter 4. The terminal reliability and total number of redundant paths for all the tag values of family of GINs [41–56] and 4DMIN are calculated by this efficient methodology, which first enumerate minimal path sets and then evaluate reliability expression in compact form of sum of disjoint products (SDP) using improved multi-variable inversion (MVI) algorithm. The SDP application results are shown in Table 6.5. Terminal reliability for different tag values is computed by generating all possible paths for each source and destination pair of the network and then evaluating terminal reliability from the minimal paths using SDP. We assumed different switch reliability values (0.9, 0.8, 0.7, 0.6, and 0.5) for calculation and comparison of terminal reliability. Terminal reliability values, evaluated for various MINs, are shown in Table 6.6–Table 6.13 for different tag values.

Discussion of reliability and fault tolerance on MINs based on Global or all terminal reliability (ATR) and Broadcast reliability of these MINs are presented in Chapter 5. We have extended the traditional reliability evaluation to include a new reliability measure as multi-source multi-terminal reliability (MSMT), discussed in Chapter 5.

GIN [41] provides only one path when source and destination switch sequence number is same but it provides multiple paths for other S-D pairs as shown in Table 6.2. It provides different level of fault tolerance to different switch pairs but fails to guarantee high fault tolerance for all source destination node pairs.

By adding an extra stage, ESGIN [42] provides more redundant paths and improve the fault tolerance capability. The advantage of 2 disjoint paths is nullified by requirement of extra switch in each path (5 stages in ESGIN as compare to 4stages in other GINs).

Table 6.5 Terminal reliability expression of 8×8 4DMIN.

Tag Value (δ)	Number of paths	Terminal reliability
0 to 7	6	$6r^4 - 4r^5 - 9r^6 + 12\, r^7 - 4r^8$

Table 6.6 Terminal reliability of the gamma network for the tag value-0.

	Switching elements reliability				
Network	**0.90**	**0.80**	**0.70**	**0.60**	**0.50**
GIN	0.6561	0.4096	0.2401	0.1296	0.0625
ESGIN	0.8048	0.6084	0.4152	0.2468	0.1226
PCGIN	0.7844	0.5587	0.3564	0.2006	0.0972
FCGIN	0.7892	0.5674	0.3643	0.2055	0.0994
CGIN	0.8069	0.6219	0.4470	0.2921	0.1679
3D-CGIN	0.8082	0.6283	0.4586	0.3052	0.1782
3DGIN	0.8082	0.6284	0.4594	0.3074	0.1816
4DMIN	**0.8093**	**0.6330**	**0.4671**	**0.3154**	**0.1875**

However, the extra stage increases the failure probability, delay and cost requirements for the network.

Terminal reliability values of CGIN [46], PCGIN [47], and FCGIN [47] are found higher compared to GIN and ESGIN due to

Table 6.7 Terminal reliability of the gamma network for the tag value-1.

	Switching elements reliability				
Network	**0.90**	**0.80**	**0.70**	**0.60**	**0.50**
GIN	0.7873	0.5734	0.3841	0.2332	0.1250
ESGIN	0.7979	0.5916	0.3950	0.2300	0.1123
PCGIN	0.7986	0.5960	0.4063	0.2480	0.1320
FCGIN	0.7791	0.5399	0.3299	0.1769	0.0818
CGIN	0.8003	0.6055	0.4254	0.2714	0.1523
3D-CGIN	0.8069	0.6221	0.4470	0.2912	0.1660
3DGIN	0.8082	0.6284	0.4594	0.3074	0.1816
4DMIN	**0.8093**	**0.6330**	**0.4671**	**0.3154**	**0.1875**

Table 6.8 Terminal reliability of the gamma network for the tag value-2.

	Switching elements reliability				
Network	**0.90**	**0.80**	**0.70**	**0.60**	**0.50**
GIN	0.7217	0.4915	0.3121	0.1814	0.0937
ESGIN	0.8062	0.6156	0.4269	0.2593	0.1317
PCGIN	0.7910	0.5757	0.3792	0.2225	0.1136
FCGIN	0.7869	0.5593	0.3534	0.1959	0.0934
CGIN	0.8056	0.6160	0.4360	0.2789	0.1562
3D-CGIN	0.8087	0.6300	0.4607	0.3061	0.1775
3DGIN	0.8082	0.6284	0.4594	0.3074	0.1816
4DMIN	**0.8093**	**0.6330**	**0.4671**	**0.3154**	**0.1875**

Table 6.9 Terminal reliability of the gamma network for the tag value-3.

	Switching elements reliability				
Network	**0.90**	**0.80**	**0.70**	**0.60**	**0.50**
GIN	0.7873	0.5734	0.3841	0.2332	0.1250
ESGIN	0.7979	0.5916	0.3950	0.2300	0.1123
PCGIN	0.7987	0.5968	0.4079	0.2499	0.1335
FCGIN	0.7818	0.5489	0.3422	0.1876	0.0888
CGIN	0.7807	0.5570	0.3625	0.2125	0.1093
3D-CGIN	0.8050	0.6124	0.4281	0.2677	0.1445
3DGIN	0.8075	0.6246	0.4516	0.2971	0.1719
4DMIN	**0.8093**	**0.6330**	**0.4671**	**0.3154**	**0.1875**

availability of two disjoint and other redundant paths. The redundant paths for CGIN, PCGIN, and FCGIN are 5, 18 and 64 respectively. Even though PCGIN and FCGIN provide more redundant paths, their reliability value is lower than CGIN.

Table 6.10 Terminal reliability of the gamma network for the tag value-4.

	Switching elements reliability				
Network	**0.90**	**0.80**	**0.70**	**0.60**	**0.50**
GIN	0.6561	0.4096	0.2401	0.1296	0.0625
ESGIN	0.8062	0.6156	0.4269	0.2593	0.1317
PCGIN	0.7844	0.5587	0.3564	0.2006	0.0972
FCGIN	0.7892	0.5674	0.3643	0.2055	0.0994
CGIN	0.7807	0.5570	0.3625	0.2125	0.1093
3D-CGIN	0.8056	0.6154	0.4333	0.2733	0.1489
3DGIN	0.8082	0.6284	0.4594	0.3074	0.1796
4DMIN	**0.8093**	**0.6330**	**0.4671**	**0.3154**	**0.1875**

Table 6.11 Terminal reliability of the gamma network for the tag value-5.

	Switching elements reliability				
Network	**0.90**	**0.80**	**0.70**	**0.60**	**0.50**
GIN	0.7873	0.5734	0.3841	0.2332	0.1250
ESGIN	0.7979	0.5916	0.3950	0.2300	0.1123
PCGIN	0.7986	0.5960	0.4063	0.2480	0.1320
FCGIN	0.7791	0.5399	0.3299	0.1769	0.0818
CGIN	0.7807	0.5570	0.3625	0.2125	0.1093
3D-CGIN	0.8050	0.6124	0.4281	0.2677	0.1445
3DGIN	0.8057	0.6246	0.4516	0.2971	0.1719
4DMIN	**0.8093**	**0.6330**	**0.4671**	**0.3154**	**0.1875**

This is due to the same interconnection pattern used in CGIN, whereas chained structure is used in PCGIN (partial chain) and FCGIN (full chain). The chained structure increases the hardware cost and failure probability as any link failure breaks the chain. CGIN

Table 6.12 Terminal reliability of the gamma network for the tag value-6.

	Switching elements reliability				
Network	**0.90**	**0.80**	**0.70**	**0.60**	**0.50**
GIN	0.7217	0.4915	0.3121	0.1814	0.0937
ESGIN	0.8062	0.6156	0.4269	0.2593	0.1317
PCGIN	0.7910	0.5757	0.3792	0.2225	0.1136
FCGIN	0.7869	0.5593	0.3534	0.1959	0.0934
CGIN	0.8056	0.6160	0.4360	0.2789	0.1562
3D-CGIN	0.8087	0.6300	0.4607	0.3061	0.1775
3DGIN	0.8075	0.6246	0.4516	0.2971	0.1719
4DMIN	**0.8093**	**0.6330**	**0.4671**	**0.3154**	**0.1875**

Table 6.13 Terminal reliability of the gamma network for the tag value- 7.

	Switching elements reliability				
Network	**0.90**	**0.80**	**0.70**	**0.60**	**0.50**
GIN	0.7873	0.5734	0.3841	0.2332	0.1250
ESGIN	0.7979	0.5916	0.3950	0.2300	0.1123
PCGIN	0.7987	0.5968	0.4079	0.2499	0.1335
FCGIN	0.7818	0.5489	0.3422	0.1876	0.0888
CGIN	0.8003	0.6055	0.4254	0.2714	0.1523
3D-CGIN	0.8070	0.6221	0.4470	0.2912	0.1660
3DGIN	0.8075	0.6246	0.4516	0.2971	0.1719
4DMIN	**0.8093**	**0.6330**	**0.4671**	**0.3154**	**0.1875**

also provides three disjoint paths for tag value 0, 2 and 6 whereas PCGIN and FCGIN provide only two disjoint paths for all the tag values. So, the terminal reliability of CGIN is higher compared to PCGIN and FCGIN.

3DCGIN [53] increases the redundant paths due to availability of alternate source in the first stage between any S-D pairs compared to CGIN. Redundant paths for 3DCGIN vary from 6 to 8 for different tag values while number of redundant paths for CGIN and GIN varies from 2 to 5 and 1 to 5 respectively. Higher reliability value for 3DGINs is observed compared to GIN and CGIN.

Due to the extra fault tolerant capability and higher number of disjoint and redundant paths, various three disjoint GINs provide higher terminal reliability value in comparison with GIN, ES-GIN, and 2DGINs. Among 3DGINs, the terminal reliability value of 3DGIN [49] is slightly higher than 3DCGIN because of their switch interconnection pattern structure.

Reliability and performance of 4DMIN depend upon how effectively available disjoint paths are utilized. The higher terminal reliability value of the proposed 4DMIN for all the tag values ensure more robust performance compared to GIN, ESGIN, 2DGINs and 3DGINs architectures.

Reliable interconnection between each source and destination depends on the availability of switching nodes and links. The terminal reliability values of GIN, ESGIN, 2DGIN, 3DGIN, and proposed 4DMIN for all the tag values considering different reliability values of switches are tabulated in Table 6.6–Table 6.13. From all the results, various GINs can be ordered in decreasing terminal reliability values as follows:

4DMIN>3DGINs >2DGINs >GIN

6.4 Reliable Interconnection Network (RIN) Layout

It is observed that similar to other GINs, reliability of proposed MIN depends highly on reliability of switches in first and last stage as these stages do not have any fault tolerance. Therefore, a more fault tolerant RIN is possible if switches in all stages are highly reliable or redundancies are available at these stages as well. Thus, a future MIN design needs to have fault tolerant capability in all the stages providing more disjoint/redundant paths with suitable rerouting techniques. The design should withstand any switch failures/busy condition at less hardware cost yielding higher performance and higher reliability.

In this section, new layout design of fault tolerant MIN is discussed. This network provides 4 disjoint paths between source and destination node pair and provides dynamic rerouting between intermediate stages. Minimal paths of these RIN have been evaluated. MINs performance evaluation has been performed by calculating terminal reliability at different assumed switch reliabilities using the SDP method. Proposed RIN provides higher terminal reliability for all tag values out performing other MINs; hence it makes a perfect choice for practical interconnection networks. The proposed design can be best choice when high reliable MINs are required.

Fast routing and rerouting strategy are needed to provide reliable performance on switch/link modes. Dynamic rerouting capability is an added advantage to use available paths effectively to design a fault tolerant MIN under switch modes (busy/failed) without backtracking.

Number of disjoint paths also gives minimum number of switches need to fail to stop communication between any source and destination nodes. Providing more disjoint paths enhances MIN fault tolerance as the common elements leading to multiple path failures are absent. Therefore, it gives better information about network capability to withstand minimum number of switch failures. Providing disjoint paths between any source and destination pair increases performance and reliability of the network. Disjoint paths are commonly used in two ways. The packets can be sent from the source to the destination nodes either using one of the disjoint paths or by transmitting multiple packets simultaneously to all disjoint paths from the source to the destination nodes.

- In the former method if any switch/link fail, network could tolerate faults by backtracking approach. In backtracking, on failure of the existing switch/link, packet can find an alternate path at each stage by sending the message packet back along the traversed path through bidirectional links to take another disjoint path. It is kind

of backward recovery. It degrades system performance and increases rerouting hop count and collision rate.
- In the later approach, multiple packets are sent concurrently through all available disjoint paths. It increases the collision rate leading to poor performance.

Although the method of sending packets concurrently increases reliability in disjoint paths network, the collision rate increases quickly as system load increases.

Dynamic adaptive rerouting: Alternative approach [50, 68, 74] is to design a fault tolerant MIN under switch/link failures without backtracking. It corrects the routing tag in which the switch changes the routing path to find an alternative path once a faulty or busy element is met. When a message packet encounters any switch/link faults, they automatically change their routing path and effectively handle collision problems.

The major observations from the literature, as discussed in Chapter 2 are:

- General assumptions are commonly made in all the MIN designs such that first and last stage switches are always in working condition and highly reliable. The intermediate stage switch modes are only considered.
- Multiplexers at the initial stage and demultiplexers at the last stage are used to withstand initial and final stage switch modes. But they automatically increase MIN hardware complexity.
- The number of crossbars switches of various MIN architectures is high and they need more layout area increasing their cost and pin outs.
- Designs are difficult to withstand multiple switch modes (busy/failed) in the intermediate stages
- There is limited comprehensive reliability evaluation and comparison on various MINs.

It is important to design highly fault tolerant dynamic MIN such that a) the design is highly fault tolerant by providing more alternate and disjoint paths, b) dynamic rerouting capability to use the disjoint paths effectively, c) withstand multiple switch modes in the intermediate stages d) fast and simple routing and rerouting strategy with low rerouting hops to reduce the collision rate and e) highly reliable architecture to meet future demands.

The major contributions of this design are summarized as follows:

- New architecture design of fault tolerant interconnection network called reliable interconnection network (RIN) is proposed.
- Proposed layout is multipath multistage interconnection network (MIN) providing four disjoint paths for all the source and destination pairs.
- Proposed layout show dynamic rerouting capability which further increased network fault-tolerant capability than other MINs.
- Proposed design can withstand input and output stage switch faults.
- Proposed design can withstand multiple switch failures in the intermediate stages.
- Proposed design is highly reliable MINs. Terminal pair reliabilities of proposed design are evaluated and analyzed with other recent proposed MINs.
- Proposed design successfully satisfies a highly reliable and low-cost layout requirement. The design has been evaluated and compared with existing MINs based on the cost.

Proposed reliable interconnection network design (RIN) is a multipath IN providing four disjoint paths and full access property, i.e. it provides connection from any of its source nodes to any of its destination nodes. RIN with the redundant paths is desirable for multiprocessor systems, as it reroute packets to other available alternative paths in case of faulty/busy path achieving enhanced reliability and performance.

The design has n+1 stage from Stage 0 to Stage n. Every Stage involves N number of sources and N number of destinations where N is 2^n.

6.4.1 Topology Design

In general, there are two main approaches to improve reliability of a MIN system design: fault tolerance and fault avoidance. Fault tolerance can be achieved by creating redundancy in system components so that if a component fails, the successor components can be used. In other words, this approach does not take any action until a fault occurs, and then takes the necessary steps to deal with system failure.

Although this approach may be helpful to avoid the failure of the whole system, it can lead to reduced system performance (e.g. increase in response time) compared to normal system operation. In addition, the redundancy can lead to an increase in design complexity, cost, weight, space requirements, and so on. In contrast, fault avoidance has a prevention strategy using high-quality and high-reliable oriented system design and it is usually more affordable compared to fault tolerance methodology.

However, there is a common point between the two approaches of fault tolerance and fault avoidance. In a rational methodology, both fault tolerance and fault avoidance approaches aimed to improve the reliability of a system and fault tolerant capability.

These major considerations play a vital role in the design of proposed reliable interconnection network-1. The proposed design implemented these two approaches successfully to provide higher fault tolerance and reliability. RIN has 4 stages (Stage 0 to 3) is shown in Figure 6.6.

Each stage involves 8 sources and 8 destinations where N=8 and n=3 (2^n). The switching elements in the first (Stage 0) and last (Stage 3) stages are 2×4 and 2×2 size crossbars. Switching elements used in the intermediate stages (Stage 1 and 2) are 4×3 and 3×2 size crossbars respectively.

Stage 0:
Various recent MIN topologies provide fault tolerance and higher reliability values using multiplexers (MUX) and demultiplexers (D-MUX) in initial stage and final stage respectively. Usage of MUX and D-Mux although increase multiple paths and path redundancy

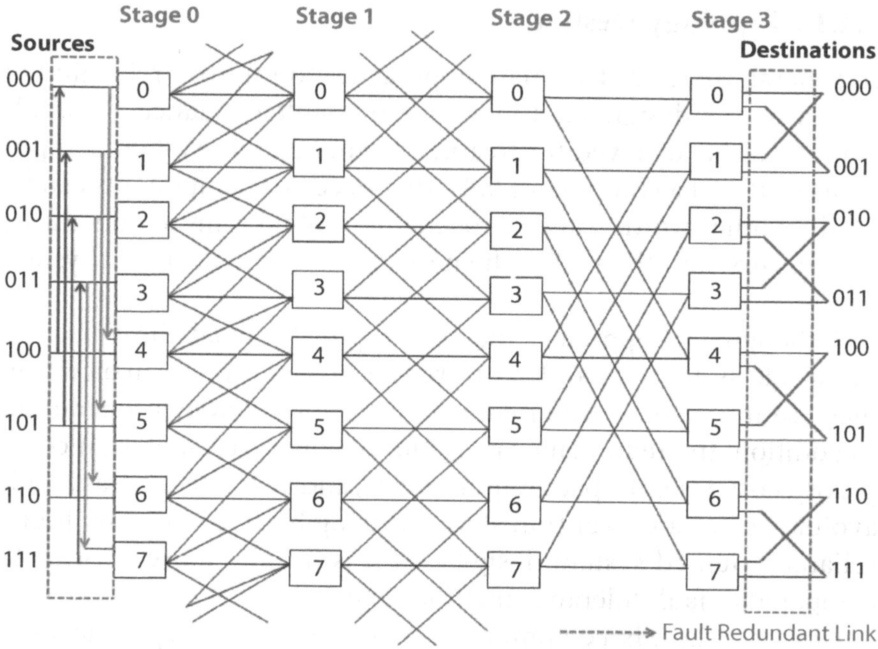

Figure 6.6 RIN showing Fault Redundant Link at Initial and Final Stage.

between node pairs but they will automatically increase their hardware complexity.

Initially, the problem formulation started with the objective of designing highly fault tolerant 4-disjoint path MIN. Therefore, switch size of Stage 0 has to be D×4 as four output links are necessary to provide four disjoint paths. Value of D has to be optimum, which can be any value out of 1, 2 and 4. Choosing D=2 makes a perfect balance where one link is input link and the second link be the fault redundant link. This fault redundant link increases number of initial stage connectivity and reduces probability of collisions. Redundant link at the initial and final stage is shown in Figure 6.6.

Stage 1 & 2:
From Stage 0, thirty-two links are emerging out. In order to evenly by-pass those links to Stage 1. We have use to use 4×D size eight crossbar switches. Other option is to use 8×D four crossbar switches.

In proposed design 4×D switches are used as 8×D crossbar switches will increase their hardware complexity.

Number of multiple paths depends largely on intermediate switch sizes. Higher number of multiple paths will be generated when D is high. A higher D provides higher redundancy, capacity and more alternate paths ensuring higher performance and reliability on switch/link failures. However, it requires complex and costly switches and can increase chances of collision or crosstalk. Lower value of D ensures simpler switch design, good link connectivity but provides lesser redundant paths resulting in lower available capacity between nodes, lower performance and lower reliability. On trying various values of D, D=3 was found to be making a balance between reliability and cost. It is observed that further increase in D to 4 and above, improves reliability very little but requires higher cost.

Therefore, eight switches of size 4×3 are used in Stage 1. Following similar logic as Stage 1 switch size selection, Stage 2 is mirror image of Stage 1 using eight 3×D crossbars. A lower value of D ensures lower dependence and simpler switch design. Choosing D=2 makes a balance and reduces required number of switches and associated costs by reducing probability of cross talk.

Stage 3:
MIN topologies provide final stage fault tolerance capability through demultiplexers (D-MUX). Usage of higher number of D-Mux however requires higher network chip size, power consumption and network cost. From Stage 2, sixteen links are emerging out. In order to evenly by-pass those links to Stage3; we have use to use 2×D size eight crossbar switches. Choosing D=2 makes a perfect balance where one link is output link and the second link be the fault redundant link. This fault redundant link increases number of output stage connectivity with the inclusion of a redundant link and faster communication between any source-destination node pairs.

6.4.2 Switching Pattern

Proposed four stage design of size 8×8 RIN is shown in Figure 6.7 and this layout provides

- Four disjoint paths between any S-D node pair.
- Twelve redundant paths for any S-D node pair.

Size of the MIN are represented by N (N=8)
Stages are represented as ***i*** (i=0,1, 2 and 3)
Switches in every Stages are represented as ***j*** (j=0, 1, 2, 3, 4, 5, 6 and 7)

At Stage 0
Number of switches be used in this stage is 8. Stage *0* is constructed by using 2×4 size crossbars. So, every switch contains four output links. These four-output links connect switches at Stage *0* and Stage *1* in the following fashion:

- First link connects ((j-2) mod N)th switch at Stage *1*
- Next link connects ((j-1) mod N)th switch at Stage *1*
- Next link connects ((j) mod N)th switch at Stage *1* and
- Fourth link connects ((j+1) mod N)th switch at Stage *1*

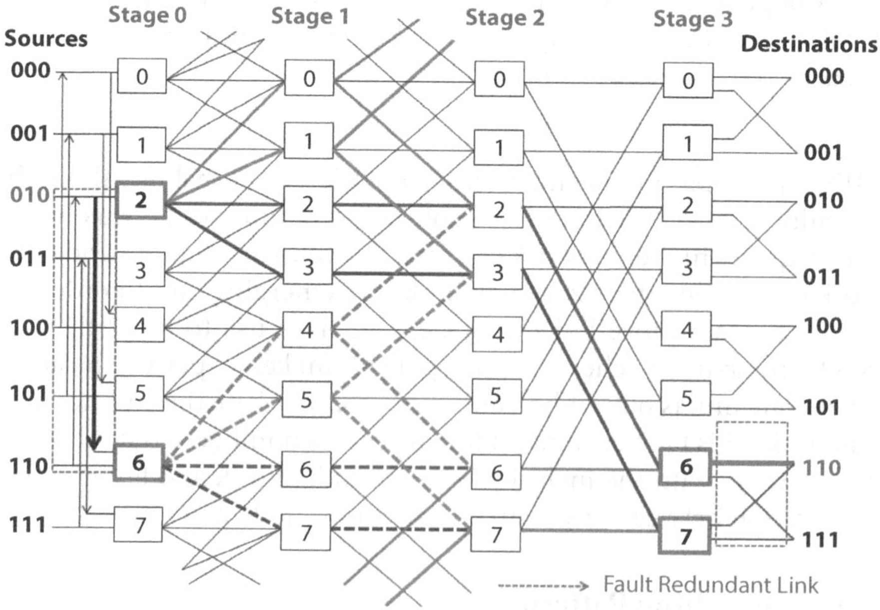

Figure 6.7 RIN showing Four Disjoint and Twelve Redundant Paths for S-D node pair (2, 6).

Fault redundant link is attached to the jth and $(j+4)^{th}$ switch at Stage 0 for initial Stage fault tolerance.

At Stage 1
Number of switches used in this stage is 8. Stage 1 is designed by using 4×3 size crossbars. So, every switch contains three output links connecting to switches at Stage 1 and Stage 2 based on the plus-minus 2^i functions. This helps in achieving dynamic rerouting capability. These three-output links connect switches at Stage *1* and Stage *2* in the following order:

- First link connects $[(j-2^i) \bmod N]^{th}$ switch at Stage *2*
- Next link connects j^{th} switch at Stage *2* and
- Final link connects $[(j+2^i) \bmod N]^{th}$ switch at Stage *2.*

At Stage 2:
Number of switches used in this stage is 8. Stage *2* is designed by using 3×2 size crossbars. So every switch contains only two output links which connects switches at Stage *2* and Stage *3*. At Stage 2, the jth switch has 2 output links a) straight link connecting to node j at Stage 3, b) non straight link connecting to node (j+4).

At Stage 3:
Number of switches used in this stage is 8. Stage *2* is designed by using 2×2 size crossbars. Fault redundant link is attached to every j^{th} and $(j+1)^{th}$ switch for final stage fault tolerance.

6.4.3 Routing and Dynamic Rerouting

Destination-tag routing is used in RIN. In destination-tag routing, message routing is carried out through destination address. Here switch settings are determined solely by the message destination address $(d_0 d_1 d_2 \ldots d_{n-1})$. RIN uses destination-based routing method to route packets. Outgoing links and the destination bits representation are shown in the Figure 6.8.

Switch/link selection for 8×8 RIN based on:

- Stage 0 decisions rely on $d_0 d_1$ bits
  Four outgoing links with the two destination bits are shown in the Figure 6.8.
- Stage 1 selection based on $d_1 d_2$ bits
  Three outgoing links with the two destination bits are shown.
- Stage 2 selections through d_2 bits
  Two outgoing links with one destination bit is shown.

Illustration:
Consider Figure 6.9 in which message packets are routed from source node 0 (000) to destination node 7 (111) through various stages with the help of destination bit address (111).

- Stage 0 decisions rely on $d_0 d_1$(11) bits
  Fourth outgoing link with destination bit (11) is selected.

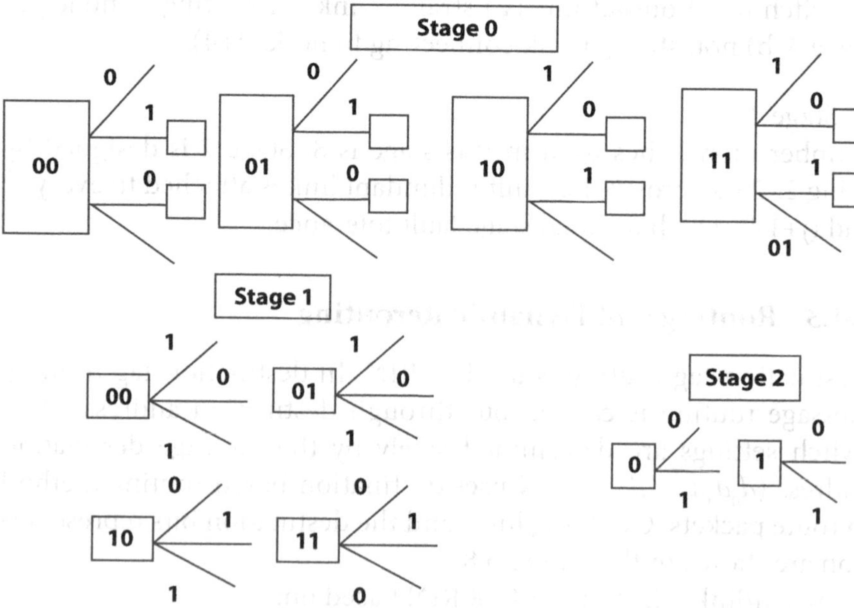

Figure 6.8 Outgoing Links and the Routing Bits from Stage 0 to Stage 2 depending on Destination Bits.

- Stage 1 selection based on $d_1d_2(11)$ bits
  Third outgoing link with destination bit (11) is selected.
- Stage 2 selections through $d_2(1)$ bits
  Non straight link with one destination bit (1) is selected.

Notations:

j = Number of Switches in a Stage i

k = Binary representation of switch number at the Stage i.

d = Destination address bits

N= Number of Source-Destination nodes (N = 8) Where n = $\log_2$N

i = Number of Switching Stages where i = (n+1) Where i = 0, 1, 2 and 3

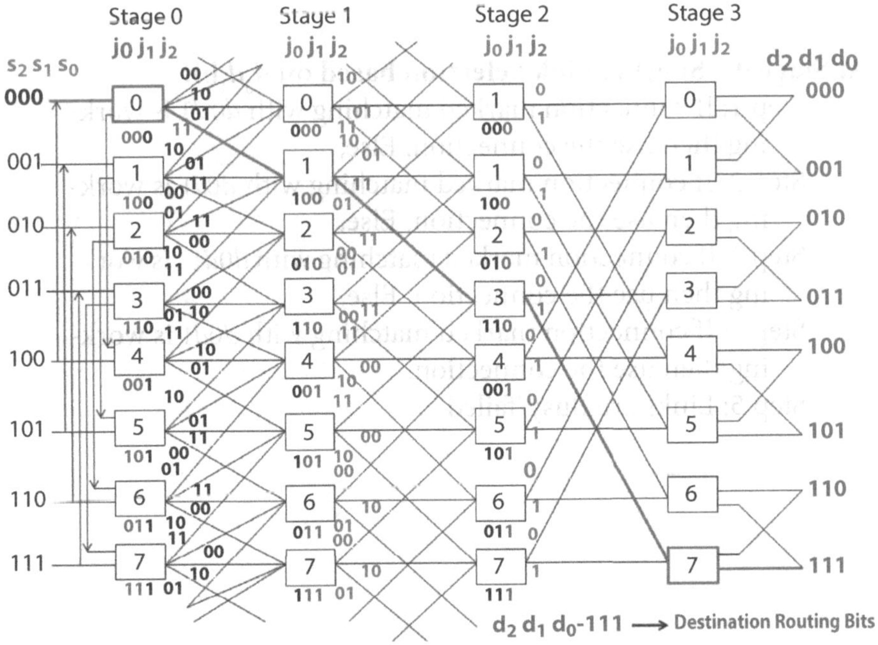

Figure 6.9 Transmission path for S-D node pair (0, 7) from Stage 0 to Stage 3 depending on Destination Tag Routing Bits.

- $j = j_0\, j_1 \ldots\ldots\ldots j_{n-1}$ be the switches at Stages 1-3. Thus $j_0\, j_1 \ldots\ldots\ldots j_7$ be the switches at Stages 1 and 2 and $j_0\, j_1\, j_2\, j_3$ switches at Stages 3.
- Destination bits are represented as $d = d_0 d_1 d_2 \ldots d_{n-1}$.
- $k_0 k_1 k_2 \ldots k_{n-1}$ be the binary representation of switch number at the Stage i. Switch number **3** be represented as 110 and Switch number 4 be 001.

Inputs
Destination address $S = s_0\, s_1\, s_2$ and $D = d_0\, d_1\, d_2$, where s_0/d_0 is the least and s_2/d_2 is the most is significant digit of destination address (0-7) in binary form.

At Stage 0
Step 1: If the switch matching source address at Stage 0 is working then use the connection. Else,
Step 2: Use the alternate switch connecting through the fault redundant link

At Stage 0 – Stage 1: Link Selection based on $d_0 d_1$ bit
Step 1: If connection marked matching with $d0d1$ is working then use the connection. Else,
Step 2: If connection marked matching with $\overline{d0}d1$ is working then use the connection. Else,
Step 3: If connection marked matching with $d0\overline{d1}$ is working then use the connection. Else,
Step 4: If connection marked matching with $\overline{d0d1}$ is working then use the connection.
Step 5: Links are Busy/failed.

$$\left(j_0j_1, d_0d_1\right) = \begin{cases} & \text{if } j_0j_1 = 00 \text{ and } d_0d_1 = 00, \text{ if } j_0j_1 = 10 \text{ and } d_0d_1 = 10 \\ & \text{if } j_0j_1 = 01 \text{ and } d_0d_1 = 01, \text{ if } j_0j_1 = 11 \text{ and } d_0d_1 = 11 \\ (j-2) \bmod N & \text{if } j_0j_1 = 00 \text{ and } d_0d_1 = 10, \text{ if } j_0j_1 = 10 \text{ and } d_0d_1 = 01 \\ (j-1) \bmod N & \text{if } j_0j_1 = 01 \text{ and } d_0d_1 = 11, \text{ if } j_0j_1 = 11 \text{ and } d_0d_1 = 00 \\ j \bmod N & \text{if } j_0j_1 = 00 \text{ and } d_0d_1 = 01, \text{ if } j_0j_1 = 10 \text{ and } d_0d_1 = 11 \\ (j+1) \bmod N & \text{if } j_0j_1 = 01 \text{ and } d_0d_1 = 00, \text{ if } j_0j_1 = 11 \text{ and } d_0d_1 = 10 \\ & \text{if } j_0j_1 = 00 \text{ and } d_0d_1 = 11, \text{ if } j_0j_1 = 10 \text{ and } d_0d_1 = 00 \\ & \text{if } j_0j_1 = 01 \text{ and } d_0d_1 = 10, \text{ if } j_0j_1 = 11 \text{ and } d_0d_1 = 01 \end{cases}$$

At Stage 1 – Stage 2: Link Selection based on d_1d_2 bits

Step 1: If connection marked matching with $d1d2$ is working then use the connection. Else, __

Step 2: If connection matching with $d1d2$ is available then use the connection. Else,

Step 3: The connection cannot be made.

$$\left(j_ij_{i+1}, d_id_{i+1}\right) = \begin{cases} & \text{if } j_ij_{i+1} = 00 \text{ and } d_id_{i+1} = 10, \text{ if } j_ij_{i+1} = 10 \text{ and } d_id_{i+1} = 01 \\ (j-2^i) \bmod N & \text{if } j_ij_{i+1} = 01 \text{ and } d_id_{i+1} = 11, \text{ if } j_ij_{i+1} = 11 \text{ and } d_id_{i+1} = 00 \\ j \bmod N & \text{if } j_ij_{i+1} = 00 \text{ and } d_id_{i+1} = 01, \text{ if } j_ij_{i+1} = 10 \text{ and } d_id_{i+1} = 11 \\ (j+2^i) \bmod N & \text{if } j_ij_{i+1} = 01 \text{ and } d_id_{i+1} = 00, \text{ if } j_ij_{i+1} = 11 \text{ and } d_id_{i+1} = 10 \\ & \text{if } j_ij_{i+1} = 00 \text{ and } d_id_{i+1} = 11, \text{ if } j_ij_{i+1} = 10 \text{ and } d_id_{i+1} = 00 \\ & \text{if } j_ij_{i+1} = 01 \text{ and } d_id_{i+1} = 10, \text{ if } j_ij_{i+1} = 11 \text{ and } d_id_{i+1} = 01 \end{cases}$$

$$\left(j_1 j_2, d_1 d_2\right) = \begin{cases} \left(j-2^i\right) \bmod N & \text{if } j_0 j_1 = 00 \text{ and } d_0 d_1 = 10, \text{ if } j_0 j_1 = 10 \text{ and } d_0 d_1 = 01 \\ & \text{if } j_0 j_1 = 01 \text{ and } d_0 d_1 = 11, \text{ if } j_0 j_1 = 11 \text{ and } d_0 d_1 = 00 \\ j \bmod N & \text{if } j_0 j_1 = 00 \text{ and } d_0 d_1 = 01, \text{ if } j_0 j_1 = 10 \text{ and } d_0 d_1 = 11 \\ & \text{if } j_0 j_1 = 01 \text{ and } d_0 d_1 = 00, \text{ if } j_0 j_1 = 11 \text{ and } d_0 d_1 = 10 \\ \left(j+2^i\right) \bmod N & \text{if } j_0 j_1 = 00 \text{ and } d_0 d_1 = 11, \text{ if } j_0 j_1 = 10 \text{ and } d_0 d_1 = 00 \\ & \text{if } j_0 j_1 = 01 \text{ and } d_0 d_1 = 10, \text{ if } j_0 j_1 = 11 \text{ and } d_0 d_1 = 01 \end{cases}$$

At Stage 2: Link Selection based on d_2 bits

$$\left(j_{n-1}, d_{n-1}\right) = \begin{cases} \left(j-2^i\right) \bmod N \text{ or } \left(j+2^i\right) \bmod N & \text{if } j_2 = 0 \text{ and } d_2 = 1, \text{ if } j_2 = 1 \text{ and } d_2 = 0 \\ j & \text{if } j_2 = 0 \text{ and } d_2 = 0, \text{ if } j_2 = 1 \text{ and } d_2 = 1 \end{cases}$$

Here n=2

$$\left(j_2, d_2\right) = \begin{cases} \left(j-2^i\right) \bmod N \text{ or } \left(j+2^i\right) \bmod N & \text{if } j_2 = 0 \text{ and } d_2 = 1, \text{ if } j_2 = 1 \text{ and } d_2 = 0 \\ j & \text{if } j_2 = 0 \text{ and } d_2 = 0, \text{ if } j_2 = 1 \text{ and } d_2 = 1 \end{cases}$$

Consider the network size N=8 considering Source S=2 and Destination=6. So the routing tag = $d_0\ d_1\ d_2$ = **110**. The routing of RIN is shown in Figure 6.10.

Case 1: Normal Condition:

- Stage 0: $\boldsymbol{d_0 d_1}$ (01) so use the switch 0 at Stage 1.
- Stage 1: $\boldsymbol{d_1 d_2}$ (11) so use the switch 2 at Stage 2.
- Stage 2: choose the switch 6 at Stage 3 by $\boldsymbol{d_2}$(1).

Case 2: Switching Modes at Stage 2:
When switch 2 ($\boldsymbol{d_1 d_2}$(11)) is busy/failed at Stage 2, the algorithm will enable switch 0 ($\boldsymbol{d_0 d_1}$(01)) at Stage 1 to choose switch 6 $\left(d1\overline{d2}(10)\right)$ as

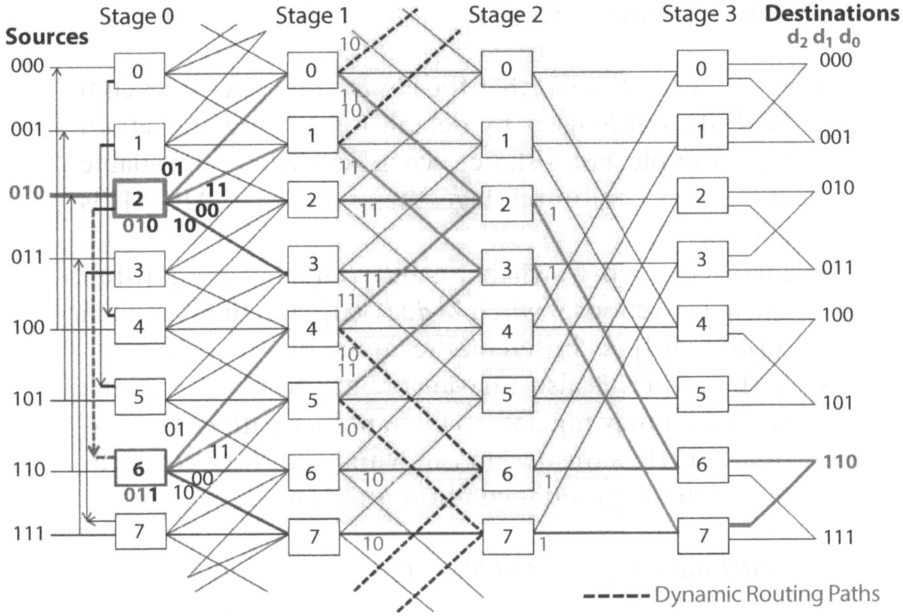

Figure 6.10 RIN Dynamic Routing Behavior for S-D node pair (2, 6). (Solid lines show the paths from source 2. Dashed lines showing the paths from source 6, if switch 2 at Stage 0 under faulty/busy condition).

the dynamic rerouting path at Stage 2 and finally switch 6 ($\boldsymbol{d}_2(1)$) at Stage 3 is chosen.

- Considering there is no message interruption at Stage 1 but switch 6 is busy/failed at Stage 2, then there is no use of switch 0 at Stage 1.
- In this condition, the algorithm automatically selects switch 1 $\left(\overline{d0}d1(11)\right)$ at Stage 1 as the next link. The next link bypasses the data *via* switch 3($\boldsymbol{d}_1\boldsymbol{d}_2(11)$) in Stage 2.
- Again, if the switch 3 is busy/failed at Stage 2 then the routing algorithm routes the data through dynamic rerouting path by switch 7 $\left(d1\overline{d2}(10)\right)$ at Stage 2.
- Even if the switch 7 is busy/failed, the algorithm will not able to send data because the design can withstand only 3 switch failures at Stage 2.

Case 3: Switch Failures at Stage 1:

- It is already known that the algorithm selects switch 0 ($d_0d_1(01)$) at Stage 1 by default for routing packets in this example. For instance, when switch 0 is unavailable at the very beginning stage, then switch 1 $\left(\overline{d0}d1(11)\right)$ is chosen.
- Likewise, even if switch 1 is busy/failed, the algorithm can able to choose switch 2 $\left(d0\overline{d1}(00)\right)$ and finally switch 3 $\left(\overline{d0}\overline{d1}(10)\right)$ is chosen on the failure of switch 2.
- If the switch 3 also unavailable at the first stage, the algorithm may not able to find another path to the destination. Thus the design can withstand three consecutive switching modes conditions at Stage 1.

Case 4: Switching Conditions at Stage 0:

- When switch 2 is busy/failed at Stage0, the algorithm will enable switch 6 at Stage 0 through the dynamic fault redundant link attached to the jth and $(j+4)^{th}$ switch at Stage 0.
- Disjoint paths and dynamic rerouting will increase faster communication and message transmission in MINs through preceding cases.

Case 1: Switch is in busy/working state
When a packet arrives at a Busy/Working switch (Switch 3 at Stage 2) dynamic rerouting node (Switch 1 at Stage 1) automatically finds an alternative path (Switch 7 at Stage 2) as shown in Figure 6.11.

Case 2: Switch is in faulty/failed state
The design can send data from any source to any destination when they meet 7 switch failures at Stage 1 and 5 switch failures at Stage 2 (except the switches connecting to the working switch of Stage 1) shown in the Figure 6.11.

Proposed reliable interconnection network design (RIN) is a simple switching element design over other architectures which provides

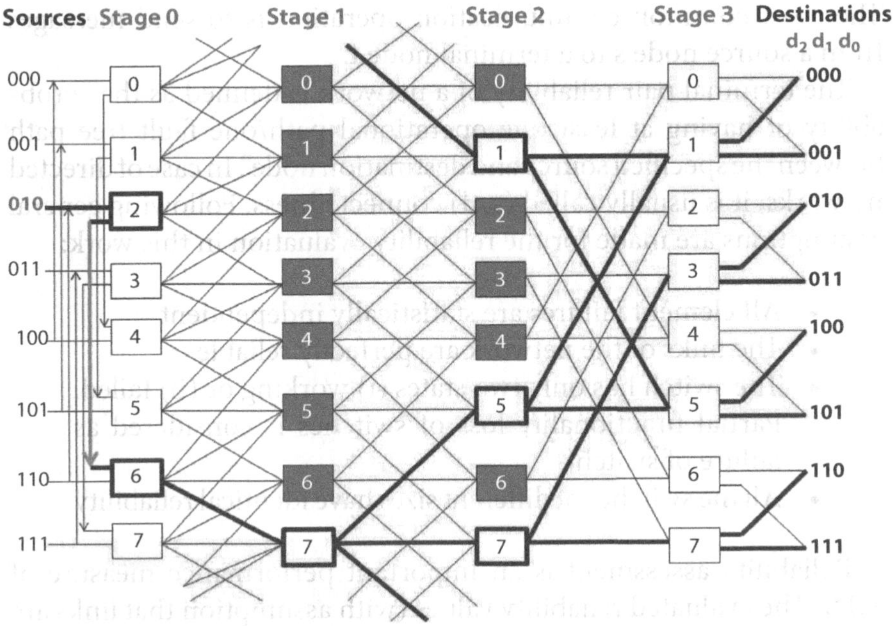

Figure 6.11 RIN showing paths from Source 2 to all the Destinations when switches at Intermediate Stages are in faulty/busy condition.

- Four disjoint paths for every source - destination node pair
- Higher Redundant (Twelve) paths for every S-D node pairs
- Dynamic rerouting capability in case of switch/link modes.
- Both initial and final stage fault tolerance.
- Higher reliability values than other MIN architectures are the major achievements of proposed design.

6.5 Reliability Analysis and Comparison of MINs

In this section, reliability of various recent topologies of MINs is analyzed. Evaluation of terminal reliability measures (indices) has attracted a lot of attention from researchers and many approaches.

The most common communication operation is to send messages from a source node s to a terminal node t.

The terminal pair reliability of a network is defined as the "probability of having at least one operational path/one fault free path between the specified source and destination node". In case of directed networks, it is usually called (s, d) connectedness. Following general assumptions are made for the reliability evaluation in this work:

- All element failures are statistically independent
- The links of the network are perfectly reliable
- The switch has only two states (i) working or (ii) failed. Partial functionality loss of switches is considered as failure of switch.
- All the switches of different sizes have identical reliability

Reliability assessment is an important performance measure of MIN. The evaluated reliability values (with assumption that links are perfectly reliable and for different switch reliability values) are the maximum achievable values considering that re-routing is capable of exploiting all available alternate paths in case of failures.

The terminal reliability and total number of redundant disjoint paths for all the MINs are calculated by enumerating the path sets using path set method and evaluating reliability expression is using hybrid multiple variable inversion (MVI) algorithm as discussed in Chapter 4. This method includes:

- Finding path sets for specified pair of nodes.
- Arranging the path sets of increasing order of cardinality
- Evaluating terminal pair reliability from ordered minimal path sets using MVI algorithm.

Terminal reliability of some recent MINs such as generalized SEN [60], tandem omega [62], extended generalized SEN [63], Clos network [64], modified baseline [65], scalable crossbar [66], extra-stage Benes [67], parallel Benes [67], replicated Benes [67], improved extra group (IEGN) [68] and pars network [69] are compared with our reliable interconnection network topologies.

The reliability of MINs is computed by generating all possible paths for each source and destination pair of the network and then evaluating reliability from the paths Considering three switch reliability values are shown in Table 6.14. Proposed MINs provides higher reliability values compared to other MIN architectures. The higher reliability values of the proposed RINs ensure more robust performance compared to other MIN architectures. Number of redundant paths and disjoint paths for MIN architectures are also presented.

Hardware complexity of multiple-path MINs (Section 2.7) is more than unique-path MINs. Hardware complexity of the design mainly depends on the number and sizes of switches used, multiplexers and de- multiplexers. Among multiple-path MINs ([59] and [67–69]) used Mux at the initial stage and D-Mux at the last stage increased their cost.

The reliability values of various MINs are evaluated and compared with proposed designs. Two-terminal reliability values of RIN are found to be higher than all 4 disjoint MINs, 3 disjoint MINs and 2 disjoint MINs. RIN is a high reliable network outperforms other single tolerant networks and traditional MINs.

Replicated Benes [67] provides two disjoint paths and high reliability values comparing other double tolerant networks Pars network [69] achieved significant tolerance than other three disjoint networks using multiplexers (MUX) and demultiplexers (D-MUX) at initial and final stages.

Clos [64] network based on a dynamic probing mechanism having rearrangeable property with a modest number of middle-stage switches to minimize implementation cost achieves higher tolerance. RIN provides higher reliability values than other 3 disjoint and 2 disjoint MINs such as extra-stage Benes [67], parallel Benes [67], replicated Benes [67], improved extra group (IEGN) [68].

Other MINs ([60, 62, 63, 65]) are notable high-fault tolerant topologies with some modern design ideas. Proposed RIN are highly fault tolerant network with higher reliability values using modest layout.

Table 6.14 Multistageinterconnection networks terminal reliability evaluation.

			TPR for Switch Reliability (r) =				
Network types	**Paths**	**Disjoint paths**	**0.95**	**0.90**	**0.80**	**0.70**	**0.60**
Modified Baseline [65]	1	1	0.8145	0.6561	0.4096	0.2401	0.1296
Parallel Benes [67]	16	2	0.9799	0.8933	0.7766	0.5331	0.3140
Tandem Omega [62]	32	2	0.8933	0.7766	0.5331	0.3140	0.1538
Extra-Stage Benes [67]	8	2	0.8931	0.7755	0.5323	0.3169	0.1596
Generalized SEN [60]	2	2	0.8940	0.7807	0.5570	0.3626	0.2125
Replicated Benes [67]	8	2	0.9886	0.9508	0.7928	0.5638	0.3354
Improved extra group [68]	12	3	0.9006	0.7971	0.5686	0.3424	0.1675
Pars Network [69]	6	3	0.9006	0.7968	0.5684	0.3447	0.1725
Extended GSEN [63]	8	4	0.9025	0.8089	0.6203	0.4173	0.2281
Clos Network [64]	4	4	0.9025	0.8099	0.6390	0.4860	0.3508
4DMIN	6	4	0.9024	0.8093	0.6330	0.4671	0.3154
Reliable IN (RIN)	**12**	**4**	**0.9998**	**0.9776**	**0.8990**	**0.7560**	**0.5634**

6.6 Summary

In this chapter two new layout designs of fault tolerant reliable interconnection networks are proposed. Both designs provide 4 disjoint paths for each source-destination node pair at each stage and dynamic rerouting capability to use the various available paths effectively. The proposed designs can withstand switch failures in all the stages and provide more redundant paths to achieve higher network reliability.

Proposed RINs are designed with highly adaptive reliable multipath dynamic routing behavior and fault tolerant capability in all the stages with suitable routing techniques. The designs also yield higher reliability. The results of performance analyses in terms of terminal reliability and hardware cost suggest that RIN surpasses other MINs which are widely used in practical interconnection networks owing to their uncomplicated configuration. RINs will act as a reliable switching architecture for designing high speed packet switches.

References

1. Frank, H. and Frisch, I.T., Analysis and design of survivable networks. *IEEE Trans. Commun.*, 18, 501–519, 1970.
2. Wilkov, R.S., Analysis and design of reliable computer networks. *IEEE Trans. Commun.*, 20, 660–678, 1972.
3. Colbourn, C.J., *The combinatorics of Network Reliability*, Oxford University Press, New York, 1987.
4. Misra, K.B. and Rao, T.S.M., Reliability analysis of redundant networks using flow graphs. *IEEE Trans. Reliab.*, 19, 19–24, 1970.
5. Misra, K.B., *Reliability Analysis and Prediction A Methodology Oriented Treatment*, Elsevier, Amsterdam, the Netherlands, 1992.
6. Misra, K.B. (Ed.), *New Trends in System Reliability Evaluation*, Elsevier, Amsterdam, the Netherlands, 1993.
7. TOP500 Supercomputer sites, http//www.top500.org/, 2020.
8. Clark, S., Haselhorst, K., Imming, K., Irish, J., Krolak, D., Ozguner, T., Cell broadband engine interconnect and memory interface, in: *Hot Chips 17*, CA, Palo Alto, August 2005.
9. Jha, S.K. and Jana, P.K., *Study and design of parallel algorithms for interconnection networks*, LAP Lambert Academic Publishing, Germany, 2011.
10. Dally, W.J. and Towles, B.P., *Principles and practices of interconnection networks*, Morgan Kaufmann Publishers, San Francisco, 2004.
11. Chaki, N. and Bhattacharya, S., Performance analysis of multistage interconnection networks with a new high-level net model. *J. Syst. Archit.*, 52, 56–70, 2006.
12. Duato, J., Yalamanchili, S., Ni, L.M., *Interconnection networks an engineering approach*, Morgan Kaufmann Publishers, San Francisco, 2003.
13. Liu, V.W., Chen, C.Y., Chen, R.B., Optimal All-to-All personalized exchange in d-nary banyan multistage interconnection networks. *J. Comb. Optim.*, 14, 131–142, 2007.
14. Chou, W.Y. and Chen, C.Y., All-to-All personalized exchange in generalized shuffle-exchange networks. *Theor. Comput. Sci.*, 411, 1669–1684, 2010.

15. Arabnia, H.R. and Smith, J.W., A reconfigurable interconnection network for imaging operations and its implementation using a multistage switching box, in: *Proceedings of seventh Annual International High Performance Computing Conference*, pp. 349–357, 1993.
16. Bhandarkar, S.M. and Arabnia, H.R., The REFINE multiprocessor — theoretical properties and algorithms. *Parallel Comput.*, 21, 1783–1805, 1995.
17. Lan, J.K., Chou, W.Y., Chen, C.Y., Efficient routing algorithms for generalized shuffle-exchange networks. *Discrete Math. Algorithms Appl.*, 1, 267–281, 2009.
18. Sharma, S., Agarwal, P., Jena, S.K., Energy aware multipath routing protocol for wireless sensor networks, in: *Computer Networks & Communications (NetCom), Lecture Notes in Electrical Engineering*, vol. 131, pp. 753–760, Springer, New York, 2013.
19. Sharma, S., *On a Class of multistage interconnection networks in parallel processing*, Ph.D. dissertation, Guru Nanak Dev University, India, 2010.
20. Dietmar, T. and Marus, B., MINSimulte - A multistage interconnection network simulator, in: *Proceedings of European Simulation Multiconference*, pp. 211–216, 2003.
21. Dietmar, T., Daniel, L., Arvid, W., Marus, B., CINSIM - A component based interconnection network simulator for modeling dynamic reconfiguration, in: *Proceedings of twelveth International conference ASMTA*, pp. 32–39, 2005.
22. Veglis, A.A., Barbargires, C.A., Pombortsis, A.S., Teaching performance evaluation of multiprocessor architectures with MathCAD and MathConnex. *IEEE Trans. Educ.*, 45, 231–237, 2002.
23. Gunawan, I., Reliability analysis of shuffle-exchange network systems. *Reliab. Eng. Syst. Saf.*, 93, 271–276, 2008.
24. Banerjee, A., Gavrilas, M., Ivanov, O., Chattopadhyay, S., Reliability improvement and the importance of power consumption optimization in wireless sensor networks, in: *Proceedings of ninth International Symposium on Advanced Topics in Electrical Engineering*, pp. 735–740, 2015.
25. Patro, A., Biswas, S., Goswami, D., Use of reliability metrics to compare bit torrent and network coding, in: *Proceedings of International Conference on Communication, Computers and Devices*, 2010.
26. Vijayalakshmi, S., Anpalagan, A., Kothari, D.P., Woungang, I., Obaidat, M.S., A comparative simulation study on the power performance of multiCore architecture. *J. Supercomputing*, 70, 465–487, 2014.

27. Bauer, E., *Design for reliability information and computer based systems*, Wiley, Hoboken, 2010.
28. Koren, I. and Krishna, C.M., *Fault-tolerant systems*, Morgan Kaufmann, San Francisco, 2007.
29. Dubois, M., Annavaram, M., Stenstrom, P., *Parallel computer organization and design*, Cambridge University Press, New York, 2012.
30. Shooman, M.L., *Reliability of computer systems and networks fault tolerance, analysis and design*, Wiley Interscience - IEEE Press, New York, 2001.
31. Jiang, X., Pattavina, A., Horiguchi, S., Strictly nonblocking fCast photonic networks. *IEEE/ACM Trans. Networking*, 16, 732–745, 2008.
32. Yang, Y. and Wang, J., Routing permutations on baseline networks with node-disjoint paths. *IEEE Trans. Parallel Distrib. Syst.*, 16, 737–746, 2005.
33. He, R. and Delgado-Frias, J.G., Fault tolerant interleaved switching fabrics for scalable high performance routers. *IEEE Trans. Parallel Distrib. Syst.*, 18, 1727–1739, 2007.
34. Fan, C.C. and Bruck, J., Tolerating multiple faults in multistage interconnection networks with minimal extra stages. *IEEE Trans. Comp.*, 49, 998–1004, 2000.
35. Shen, X., Yang, F., Pan, Y., Equivalent permutation capabilities between time-division optical omega networks and non-optical extra-stage omega networks. *IEEE Transactions on Networking*, 9, 4, 518–524, 2001.
36. Kumar, V. and Reddy, S., Augmented shuffle-exchange multistage interconnection networks. *IEEE Trans. Comp.*, 20, 30–40, 1987.
37. Nieminen, E., A contention-free parallel access by butterfly networks for turbo interleavers. *IEEE Trans. Inf. Theory*, 60, 237–251, 2014.
38. Bansal, P., Singh, K., Joshi, R.C., Quad tree a cost-effective fault-tolerant multistage interconnection network, in: *Proceedings of Eleventh Annual Joint Conference of the IEEE Computer and Communications Societies*, pp. 860–866, 1992.
39. Sengupta, J. and Bansal, P., High speed dynamic fault-tolerance, in: *Proceedings of IEEE Region 10 International Conference on Electrical and Electronic Technology*, pp. 669–675, 2001.
40. Sengupta, J. and Bansal, P., Performance analysis of static and dynamic fault-tolerant irregular networks, in: *Proceedings of IEEE Region International Conference on Electrical and Electronic Technology*, pp. 5–8, 2004.
41. Parker, D. and Raghavendra, C., The gamma network. *IEEE Trans. Comp.*, 100, 367–373, 1984.

42. Lee, K.Y. and Hegazy, W., The extra stage gamma network. *IEEE Trans. Comp.*, 37, 1445–1450, 1998.
43. Lee, K.Y. and Yoon, H., The B-network A multistage interconnection network with backward links. *IEEE Trans. Comp.*, 39, 7, 966–969, 1990.
44. Chuang, P.J., CGIN A modified gamma interconnection network with multiple disjoint path, in: *Proceedings of International Conference of Parallel and Distributed Systems*, pp. 366–372, 1994.
45. Tzeng, N.F., Chuang, P.J., Wu, C.H., Creating disjoint paths in gamma interconnection networks. *IEEE Trans. Comp.*, 42, 1247–1252, 1996.
46. Chuang, P.J., CGIN a fault tolerant modified gamma interconnection network. *IEEE Trans. Parallel Distrib. Syst.*, 7, 1301–1306, 1996.
47. Chen, C.W., Lu, N.P., Chen, T.F., Chung, C.P., Fault-tolerant gamma interconnection networks by chaining. *Proc. IEE Comput. Digital Tech.*, 147, 75–81, 2000.
48. Chen, C.W. and Chung, C.P., Fault-tolerant gamma interconnection network without backtracking. *J. Syst. Software*, 58, 23–31, 2001.
49. Chen, C.W., Lu, N.P., Chung, C.P., 3-Disjoint gamma interconnection networks. *J. Syst. Software*, 66, 2, 129–134, 2003.
50. Chen, C.W. and Fu, S.C., A minimal links traversed dynamic rerouting network. *Parallel Comput.*, 30, 883–898, 2004.
51. Chen, C.W. and Chung, C.P., Designing a disjoint paths interconnection network with fault tolerance and collision solving. *J. Supercomputing*, 34, 63–80, 2005.
52. Cheng, L., Venkatesan, R., Heys, H.M., Architecture and performance analysis of the multicast balanced gamma switch for broadband communications, in: *Proceedings of IEEE International Conference on Computer Systems and Applications*, pp. 381–388, 2006.
53. Borkar M.A., Nitin (2011) 3D-CGIN: A 3 Disjoint Paths CGIN with Alternate Source. In: *Advances in Computing and Communications. ACC 2011. Communications in Computer and Information Science.* Abraham, A., Mauri, J.L., Buford, J.F., Suzuki, J., Thampi, S.M. (eds), vol 193. Springer, Berlin, Heidelberg Kochi, India, July 22-24, 2011.
54. Rastogi, R., Verma, R., Nitin and Chauhan, D.S., 3-Disjoint paths fault-tolerant multiStage interconnection networks. *Proc. Adv. Comput. Commun.*, 190, 21–33, 2011.
55. Chen, Z., *A class of incomplete gamma interconnection network*, pp. 1–12, Technical Report - RIIT-TNLIST, Technical Report China Tsinghua University, 2013.
56. Venkatesan, R. and Mouftah, H., Balanced gamma network A new candidate for broadband packet switch architectures. *Proc. Eleventh Annual Joint Conf. IEEE Comp. Commun. Soc.*, 3, 2482–2488, 1992.

57. Raponi, P.G., Andriolli, N., Cerutti, I., Torres, D., Liboiron-Ladouceur, O., Castoldi, P., Heterogeneous optical space switches for scalable and energy-efficient data centers. *J. Lightwave Technol.*, 31, 1713–1719, 2013.
58. Yang, Y. and Wang, J., A class of multistage conference switching networks for group communication. *IEEE Trans. Parallel Distrib. Syst.*, 15, 228–243, 2004.
59. Aggarwal, R. and Kaur, L., An efficient routing scheme to provide more fault-tolerance for an irregular multistage interconnection network, in: *Proceedings of IEEE International Advance Computing Conference*, pp. 94–98, 2009.
60. Chiuyuan, C. and JingKai, L., An efficient tag Based routing algorithm for the backward network of a bidirectional general shuffle-exchange network. *IEEE Commun. Lett.*, 10, 296–298, 2006.
61. Park, J.H., Routing correctness proof and reliability analysis of the two-dimensional Ring Banyan network. AEU - *Int. J. Electron. Commun.*, 63, 968–976, 2009.
62. Li, S.Y. and Tan, X., On rearrangeability of tandem connection of banyan-type networks. *IEEE Trans. Commun.*, 57, 164–170, 2009.
63. Busi, I. and Pattavina, A., Non blocking conditions for EGS networks. *IEEE Commun. Lett.*, 14, 572–574, 2010.
64. Pham, P.H., Song, J., Park, J., Kim, C., Design and implementation of an on Chip permutation network for multiprocessor system-on Chip. *IEEE Trans. Very Large Scale Integr VLSI Syst.*, 21, 173–177, 2013.
65. Danilewicz, G. and Rajewski, R., The architecture and strict sense nonblocking conditions of a new baseline based optical switching network composed of symmetrical and asymmetrical switching elements. *IEEE Trans. Commun.*, 62, 1058–1069, 2014.
66. Bistouni, F. and Jahanshahi, M., Scalable crossbar network a non blocking interconnection network for large scale systems. *J. Supercomputing*, 71, 697–728, 2015.
67. Bistouni, F. and Jahanshahi, M., Improving the reliability of the Benes network for use in large scale systems. *Microelectron. Reliab.*, 55, 679–695, 2015.
68. Bistouni, F. and Jahanshahi, M., Improved extra group network a new fault-tolerant multistage interconnection network. *J. Supercomputing*, 69, 161–199, 2014.
69. Bistouni, F. and Jahanshahi, M., Pars network A multistage interconnection network with fault-tolerance capability. *J. Parallel Distrib. Comput.*, 75, 168–183, 2015.
70. Bistouni, F. and Jahanshahi, M., A new approach to improve reliability of the multistage interconnection networks. *Comput. Electr. Eng.*, 40, 348–374, 2014.

71. Bistouni, F. and Jahanshahi, M., Analyzing the reliability of shuffle-exchange networks using reliability block diagrams. *Reliab. Eng. Syst. Saf.*, 132, 97–106, 2014.
72. McMillen, R.J. and Siegel, H.J., Performance and fault tolerance improvements in the inverse augmented data manipulator network. *Proc. 9th Annual Symposium on Computer Architecture*, pp. 63–72, Austin, TX, USA, April 1982.
73. Clos, C., A study of non-blocking switching networks. *Bell Syst. Tech. J.*, 32, 406–424, 1953.
74. Yang, Y. and Wang, J., A new design for wide sense non blocking multicast switching networks. *IEEE Trans. Commun.*, 53, 497–504, 2005.
75. Lee, K.Y. and Yoon, H., The PM22I interconnection network. *IEEE Trans. Comp.*, 38, 902–907, 1989.
76. Cheng, X. and Ibe, O.C., Reliability of a class of multistage interconnection networks, in: *IEEE Transactions on Parallel and Distributed Systems*, vol. 3, pp. 241– 246, 1992.
77. Fard, N.S. and Gunawan, I., Terminal reliability improvement of shuffle-exchange network systems. *Int. J. Reliab. Qual. Saf. Eng.*, 12, 51–60, 2005.
78. Blake, J.T. and Trivedi, K.S., Multistage interconnection network reliability. *IEEE Trans. Comp.*, 38, 1600–1604, 1989.
79. Sien, L.C., *Extra Stage cube network reliability and algorithm*, Master Thesis, Malaysia University of Science and Technology, 2004.
80. Kett, L.W., *Shuffle exchange network with an additional stage reliability simulation*, Master Thesis, Malaysia University of Science and Technology, 2004.
81. Varma, A. and Raghavendra, C.S., Reliability analysis of redundant path interconnection networks. *IEEE Trans. Reliab.*, 38, 130–137, 1989.
82. Hwang, K. and Chang, Y., Combinatorial reliability analysis of multiprocessor computers. *IEEE Trans. Reliab.*, 31, 469–473, 1982.
83. Cherkassy, V. and Malek, M., Reliability and fail softness analysis of multistage interconnection networks. *IEEE Trans. Reliab.*, 34, 524–527, 1985.
84. Wang, W. and Mingxiao, J., Generalized decomposition method for complex systems, in: *Proceedings of Reliability and Maintainability Annual Symposium - RAMS*, pp. 12–17, 2004.
85. Gunawan, I., Reliability bounds for large multistage interconnection network, in: *Proceedings of Applied Parallel Computing Conference*, pp. 507–514, 2002.
86. Abraham, J.A., An improved algorithm for network reliability. *IEEE Trans. Reliab.*, 28, 58–61, 1979.

87. Gunawan, I., Reliability paths and reliability bounds in gamma networks. *Appl. Math. Modell.*, 32, 588– 594, 2008.
88. Tagle, P.U. and Sharma, N.K., Reliability analysis of interconnected parallel banyan network, in: *Proceedings of IEEE International Conference on communications*, pp. 651–656, 1998.
89. Tagle, P.U. and Sharma, N.K., Reliability analysis of fault-tolerant switching network for 8-ISDN, in: *Proceedings of IEEE International Conference on computers andcommunications*, pp. 165–171, 1996.
90. Tabada, L. and Tagle, P.U., Reliability analysis of a fault tolerant switch, in: *Proceedings of International Conference on Computer Engineering and Applications*, pp. 295–300, 2009.
91. Aulakh, S., Reliability analysis of mux–demux replicated multistage interconnection networks. *Exp. Tech.*, 30, 19–22, 2006.
92. Mahajan, R. and Vig, R., Reliability and performance analysis of new fault tolerant irregular network. *WSEAS Trans. Comp. Res.*, 3, 311–321, 2008.
93. Aggarwal, R.R. and Lakhwinder, K., On reliability analysis of fault-tolerant multistage interconnection networks. *Int. J. Comput. Sci. Secur.*, 2, 1–8, 2008.
94. Aggarwal, R.R. and Lakhwinder, K., Design and reliability analysis of a new fault-tolerant multistage interconnection network. *Int. J. Comp. Netw. Int. Res.*, 8, 17–23, 2009.
95. Aggarwal, R.R., Design and performance evaluation of a new irregular fault-tolerant multistage interconnection network. *Int. J. Comp. Sci.*, 9, 108–113, 2012.
96. Aggarwal, R.R., Design and reliability analysis of a class of irregular fault-tolerant multistage interconnection networks. *Int. J. Comput. Appl.*, 39, 8–14, 2012.
97. Nitin, and Subramanian, A., On reliability analysis of cost-effective hybrid zeta network a fault-tolerant multiStage interconnection network, in: *Proceedings of the international conference on parallel and distributed processing techniques and applications*, pp. 260–265, 2006.
98. Sharma, S., Kahlon, K.S., Bansal, P.K., Reliability and path length analysis of irregular fault tolerant multistage interconnection network. *ACM SIGARCH Comput. Architecture News*, 37, 5, 16–23, 2009.
99. Fishman, G.S., A comparison of four Monte Carlo methods for estimating the probability of s-t connectedness. *IEEE Trans. Reliab.*, 35, 145–155, 1986.

100. Fishman, G.S., Monte Carlo estimation of function variation. *Winter Simulation Conference Proceedings 1987,* 347–350, IEEE Service Cent (Cat n 87CH2512-2), Piscataway, NJ, USA, 1987c.
101. Fishman, G.S., *Monte Carlo - concepts, algorithms, and applications*, Springer, New York, 1996.
102. Huseby, A.B., Naustdal, M., Vårli, I.D., *System reliability evaluation using conditional Monte Carlo methods*, Statistical Research Report, University of Oslo, 2004.
103. Gunawan, I., Multistage interconnection networks reliability evaluation based on stratified sampling Monte Carlo method. *Int. J. Model. Simul.*, 28, 209–214, 2008.
104. Gunawan, I., Reliability prediction of distributed systems using Monte Carlo method. *Int. J. Reliab. Saf.*, 7, 235–248, 2013.
105. Gunawan, I. and Fard, N.S., Terminal reliability assessment of gamma and extraStage gamma networks. *Int. J. Qual. Reliab. Manage.*, 29, 820–831, 2012.
106. Aggarwal, K.K., Gupta, J.S., Misra, K.B., Reliability evaluation a comparative study of different methods. *Microelectron. Reliab.*, 14, 49–56, 1975.
107. Aziz, M.A., Soban, M.A., Samad, M.A., Reduction in computations in enumeration of terminal and multi-terminal pathset by method of indexing. *Microelectron. Reliab.*, 32, 1067–1072, 1992.
108. Aziz, M.A., Soban, M.A., Samad, M.A., Enumeration of pathsets of reliability graph by repeated indexing. *Microelectron. Reliab.*, 33, 4, 481–487, 1993.
109. Beigel, J.E., Determination of tieSets and cutSets for a system without feedback. *IEEE Trans. Reliab.*, R-26, 1, 39–42, 1977.
110. Danielson, G., On finding the simple pathsets and circuits in graph. *IEEE Trans. Circuit Theory*, 15, 294–295, 1968.
111. Fratta, L. and Montanari, U.G., A vertex elimination algorithm for enumerating all simple pathsets in a graph. *Networks*, 20, 151–177, 1975.
112. Krishnamurthy, E.V. and Komissar, G., Computer aided reliability analysis of complicated networks. *IEEE Trans. Reliab.*, 21, 86–89, 1972.
113. Misra, R.B., An algorithm for enumerating all simple paths in a communication network. *Microelectron. Reliab.*, 19, 363–366, 1979.
114. Misra, K.B., An Algorithm for the Evaluation of Redundant Network. *IEEE Trans. Reliab.*, 19, 4, 146–151, 1970.
115. Rai, S. and Aggarwal, K.K., An efficient method for reliability evaluation of general networks. *IEEE Trans. Reliab.*, 27, 206–211, 1978.

116. Samad, M.A., An efficient method for terminal and multi-terminal pathset enumeration. *Microelectron. Reliab.*, 35, 443–446, 1987.
117. Fard, N.S. and Hunhee, L., An enumeration method for the minimal pathsets of network systems. *Int. J. Reliab. Qual. Saf. Eng.*, 7, 27–41, 2000.
118. Soh, S. and Rai, S., CAREL Computer aided reliability evaluator for distributed computer networks. *IEEE Trans. Parallel Distrib. Syst.*, 2, 199–213, 1991.
119. Soh, S. and Rai, S., Experimental results on preprocessing of path/cut terms in sum of disjoint product technique. *IEEE Trans. Reliab.*, 42, 24–33, 1993.
120. Rai, S. and Agrawal, K.K., An efficient method for reliability evaluation of general networks. *IEEE Trans. Reliab.*, 27, 206–211, 1978.
121. Heidtmann, K.D., Smaller sum of disjoint product by subproduct inversion. *IEEE Trans. Reliab.*, 38, 305–311, 1989.
122. Hsu, L.H. and Lin, C.K., *Graph theory and interconnection network*, Taylor and Francis Group, CRC Press, Boca Raton, FL, 2008.
123. Yunus, N.A. and Othman, M., Reliability evaluation and routing integration in shuffle exchange omega network. *J. Networks*, 9, 1732–1737, 2014.
124. Chaturvedi, S.K. and Misra, K.B., An efficient multi-variable inversion algorithm for reliability evaluation of complex systems using path sets. *Int. J. Reliab. Qual. Saf. Eng.*, 9, 237–259, 2002.
125. Goyal, N.K., *On Some Aspects of Reliability Analysis and Design of Communication Networks*, Ph.D. dissertation, Indian Institute of Technology Kharagpur, India, 2006.
126. Rajkumar, S. and Goyal, N.K., Design of 4-disjoint gamma interconnection network layouts and reliability analysis of gamma interconnection Networks. *J. Supercomputing*, 69, 468–491, 2014.

Index

Also of Interest

Check out these published and forthcoming titles in the Performability Engineering Series

Interconnection Network Reliability Evaluation
Multistage Layouts
By Neeraj Goyal and S. Rajkumar
Published 2020. ISBN 978-1-119-62058-7

Repairable Systems Reliability Analysis
A Comprehensive Framework
By Rajiv Nandan Rai, Sanjay Kumar Chaturvedi and Nomesh Bolia
Published 2020. ISBN 978-1-119-52627-8

Probabilistic Physics of Failure Approach to Reliability: Modeling, Accelerated Testing, Prognosis and Reliability Assessment
By Mohammad Modarres, Mehdi Amiri and Christopher Jackson
Published 2017. ISBN 978-1-119-38863-0

Network Reliability
Measures and Evaluation
By Sanjay K. Chaturvedi
Published 2016. ISBN 978-1-119-22356-6

Machine Tool Reliability
By Bhupesh Kumar Lad, Divya Shrivastava and Makarand S. Kulkarni
Published 2016. ISBN 978-1-119-03860-3

Binary Decision Diagrams and Extensions for Systems Reliability Analysis
By Suprasad Amari and Liudong Xing
Published 2015. ISBN 978-1-118-54937-7

Quantitative Assessments of Distributed Systems
Methodologies and Techniques
By Dario Bruneo and Salvatore Distefano
Published 2015. ISBN 978-1-118-59521-3

Fundamentals of Reliability Engineering
By Indra Gunawan
Published 2014. ISBN 978-1-118-54956-8

Building Dependable Distributed Systems
By Wenbing Zhao
Published 2014. ISBN 978-1-118-54943-8

Printed in the USA/Agawam, MA
September 16, 2026

910661.006